INDEX OF THELYPTERIDACEAE

by

J. W. Grimes & B. S. Parris

Royal Botanic Gardens
Kew
1986

ISBN 0 947643 03 6

Phototypeset in Linotron Times by Input Typesetting Ltd London
Printed in Great Britain by Whitstable Litho Ltd Whitstable Kent

INTRODUCTION

Thelypteridaceae is one of the largest families of ferns. Many workers have tried to clarify its taxonomy (see introduction to Holttum 1969), often studying only limited amounts of material; consequently there has been much confusion.

During the last two decades monographic studies of the family in the Old World have been undertaken by Holttum, in which several genera have been recognised. In order to solve the many problems he has studied all available type specimens and has thus been able to clarify the taxonomy as well as providing usable keys and good descriptions: this treatment is certainly the most thorough and accurate that is available. A. R. Smith is undertaking similar studies with regard to the New World species.

Now that the nomenclature and taxonomy is more clearly understood, workers face problems with regard as to which are accepted specific epithets and which are synonyms, and also into which genus (or as some prefer, subgenus) a species should be placed. At present the only way to find out is the slow process of scanning all the relevant literature.

In 1968 Reed published 'Index Thelypteridis' in order to try to list all of the taxa of the Family; each species known to him is placed in the genus *Thelypteris* s.l. However, this index is far from complete and contains many inaccuracies.

To allow the rapid and efficient use of the modern taxonomic treatments we have prepared an alphabetical index of all known specific epithets which have been used in the family with the basionym, its authority and date of publication, and also the genus (subgenus) in which they are now placed, with the author and date of publication if the combination has been made. Subspecific epithets are only included when they are currently treated as belonging to species other than those under which they were originally described. This index, used in conjunction with Index Filicum and the appended bibliography, permits rapid access to references for basionyms, synonyms and accepted names.

This, as with any large index of a complex and confusing group, cannot be final; opinions will differ and there are doubtless some mistakes or omissions, however this (as far as we are able to determine) thoroughly encompasses the modern taxonomic treatments. Obviously there will be alterations and additions in the future, and as and when necessary supplements to this index will be prepared.

No new combinations (or new taxa) are proposed in this paper. All names in literature received at Kew up to the end of 1985 have been included.

We should like to thank Professor R. E. Holttum for his help and advice in the compiling and checking of this list.

THE INDEX

Epithet	Genus in which Originally Described	Present or Proposed Disposition
abbiattii	*Thelypteris* Reed (1968)	= *burkartii (Goniopteris* Abbiat.)
abbottiana	*Dryopteris* Maxon (1924)	= *Amauropelta*
abbreviata	*Goniopteris* Presl (1836)	= *glandulosum* Desv.
abbreviatipinna	*Dryopteris* Makino & Ogata (1929)	= *cystopteroides*
abdita	*Thelypteris* Proctor (1985)	= *Goniopteris*
abortivum	*Aspidium* Bl. (1828)	= *truncatum* Poir.
abruptum	*Aspidium* Bl. (1828)	= *truncatum* Poir.
abruptum	*Aspidium* Kze (1834)	= *grandis*
abruptum	*Polypodium* Desv. (1827)	= *Goniopteris*
acanthocarpa	*Dryopteris* Copel. (1911)	= *Pronephrium* Holtt. (1972)
achalense	*Aspidium* Hieron. (1896)	= *Amauropelta*
acromanes	*Dryopteris* Christ (1907)	= *diversilobum*
acrostichoides	*Dryopteris* v.A.v.R. (1917)	= *celebicum* Bak.
acrostichoides	*Nephrodium* Michx. (1803)	= *Polystichum*
acrostichoides	*Nephrodium* Desv. (1827)	*Sphaerostephanos* Holtt. (1982)
aculeata	*Thelypteris* A. R. Smith (1983)	= *Amauropelta*
acuminata	*Dryopteris* Rosenst. (1917)	= *subpubescens* Bl.
acuminatum	*Polypodium* Houtt. (1783)	*Christella* Holtt. (1976)
acuminatum	*Polypodium* Roxb. (1825)	= *aridum*
acutidens	*Chingia* Holtt. (1974)	*Chingia* Holtt. (1974)
acutum	*Polypodium* Roxb. (1844)	= *subalpina*
adenochlamys	*Dryopteris* C. Chr. (1911)	= *microbasis*
adenochrysa	*Phegopteris* Fée (1852)	= *Amauropelta*
adenopelta	*Christella* Holtt. (1976)	*Christella* Holtt. (1976)
adenophora	*Dryopteris* C. Chr. (1906)	= *hirsutum Kunze ex* Mett.
adenopteris	*Aspidium* Mett. (1858)	= *maemonensis*
adenostegia	*Dryopteris* Copel. (1942)	*Sphaerostephanos* Holtt. (1982)
adscendens	*Thelypteris* Ching (1936)	*Metathelypteris* Ching (1963)
aequatorialis	*Dryopteris* Copel. (1941)	= *Amauropelta*
aequibasis	*Dryopteris* C. Chr. (1925)	= *pulchrum*
affine	*Aspidium* Bl. (1828)	*Pronephrium* Presl (1851)
affine	*Meniscium* Ettingsh. (1864)	*Meniscium* Ettingsh. (1864)
afra	*Dryopteris* Christ (1908)	*Pneumatopteris* Holtt. (1973)
africanum	*Polypodium* Desv. (1827)	= *pozoi*
afzelii	*Dryopteris* C. Chr. (1916)	*Christella* Holtt. (1974)
alan-smithiana	*Thelypteris* L. Gómez (1982)	= *Goniopteris*
alatellum	*Nephrodium* Christ (1900)	*Sphaerostephanos* Holtt. (1982)
alatum	*Polypodium* L. (1753)	*Goniopteris* Ching (1940)
albescens	*Nephrodium* Desv. (1827)	= *patens* Sw.
albicaule	*Aspidium* Fée (1857)	= *Christella*
albidipilosa	*Dryopteris* Bonap. (1924)	= *Thelypteris* s.l.
albociliata	*Dryopteris* Copel. (1929)	= *parasiticum* L.

1

Epithet	Genus in which Originally Described	Present or Proposed Disposition
albosetosa	Dryopteris Copel. (1942)	Sphaerostephanos Holtt. (1982)
alcasidii	Sphaerostephanos Holtt. (1982)	Sphaerostephanos Holtt. (1982)
alfarii	Aspidium Christ (1905)	= pilosulum
alfredii	Dryopteris Rosenst. (1925)	= Amauropelta
aliena	Dryopteris C. Chr. (1937)	= Amauropelta
alpinus	Sphaerostephanos Holtt. (1982)	Sphaerostephanos Holtt. (1982)
alsophiloides	Phegopteris Fourn. (1873)	= torresianum
alta	Dryopteris Brause (1920)	Plesioneuron Holtt. (1975)
alticola	Sphaerostephanos Holtt. (1982)	Sphaerostephanos Holtt. (1982)
x altissima	Christella Holtt. (1974)	Christella Holtt. (1974)
amabilis	Leptogramma Tagawa (1938)	= gymnocarpa subsp.
amaiensis	Dryopteris Rosenst. (1917)	= cuspidatum Bl.
amboinense	Aspidium Willd. (1810)	Pronephrium Holtt. (1972)
amboinense	Aspidium sensu Bl., Enum. Pl. Jav. (1828)	= subpubescens Bl.
amboinense	Nephrodium Hook. (1862)	= subpubescens Bl.
var. subglandulosum	Nephrodium Bak. (1882)	= veitchii
amoyensis	cyclosorus Ching (ined)	= ? Christella
amphioxypteris	Nephrodium Sod. (1883)	= Amauropelta
amphitrichum	Pronephrium Holtt. (1982)	Pronephrium Holtt. (1982)
anateinophlebium	Nephrodium Bak. (1877)	= bergianum
anceps	Dryopteris Maxon (1922)	= guyanensis
ancyriothrix	Dryopteris Rosenst. (1909)	= Goniopteris
andersonii	Coryphopteris Holtt. (1976)	Coryphopteris Holtt. (1976)
andina	Dryopteris Morton (1938)	= Amauropelta
andreae	Coryphopteris Holtt. (1976)	Coryphopteris Holtt. (1976)
andreanum	Meniscium Sod. (1883)	Meniscium Sod. (1883)
aneitense	Aspidium Fourn. (1874)	= lenormandii
angiensis	Plesioneuron Holtt. (1982)	Plesioneuron Holtt. (1982)
angulariloba	Thelypteris Ching (1936)	= hirsutipes
angusta	Dryopteris Copel. (1914)	= acrostichoides Desv.
angusta	Syngramma Copel. (1909)	= oligodictyon
angustata	Lastrea Presl (1836)	= Amauropelta
angustibasis	Sphaerostephanos Holtt. (1982)	Sphaerostephanos Holtt. (1982)
angusticaudata	Pneumatopteris Holtt. (1973)	Pneumatopteris Holtt. (1973)
angustifolium	Meniscium Willd. (1810)	Meniscium Willd. (1810)
angustifolium	Nephrodium Presl (1851)	Sphaerostephanos Holtt. (1975)
angustifrons	Aspidium Miq. (1867)	Parathelypteris Ching (1963)
angustifrons	Trigonospora Sledge (1981)	Trigonospora Sledge (1981)
angustipes	Dryopteris Copel. (1912)	= latebrosum
angustipinnatus	Cyclosorus Tard. (1941)	= Pronephrium
anjenabensis	Abacopteris Tard. (1965)	aff. Ctenitis
anopteron	Aspidium Kze. (1869)	= Goniopteris
antillana	Thelypteris Proctor (1961)	= Amauropelta
aoristisorum	Polypodium Harr. (1897)	Nannothelypteris Holtt. (1971)

Epithet	Genus in which Originally Described	Present or Proposed Disposition
appendiculata	*Gymnogramme* Bl. (1828)	*Sphaerostephanos* Holtt. (1975)
appendiculatum	*Aspidium* Mett. (1858)	= *squamaestipes* Clarke
appendiculatum	*Nephrodium* Presl (1851)	*Christella* Holtt. (1976)
appressa	*Thelypteris* A. R. Smith (1983)	= *Amauropelta*
aquapimense	*Aspidium* Schum. (1829)	= *dentatum*
aquatilis	*Dryopteris* Copel. (1911)	*Sphaerostephanos* Holtt. (1982)
aquatiloides	*Dryopteris* Copel. (1912)	*Pronephrium* Holtt. (1972)
arborea	*Dryopteris* Brause (1914)	= *Amauropelta*
arborea	*Dryopteris* v.A.v.R. (1918)	= *arfakianum*
arborescens	*Meniscium* Willd. (1810)	= *dentatum*
arbuscula	*Aspidium* Willd. (1810)	*Sphaerostephanos* Holtt. (1974)
arbuscula	*Nephrodium* sensu Racib., Fl. Buit. (1898)	= *perglandulifera*
archboldiae	*Lastrea* Copel. (1949)	*Plesioneuron* Holtt. (1975)
archboldii	*Dryopteris* C. Chr. (1937)	*Sphaerostephanos* Holtt. (1982)
arcana	*Dryopteris* Maxon & Morton (1938)	*Meniscium* Pic. Ser. (1968)
arcuatum	*Polypodium* Poir. (1804)	= *lepida*
arechavaletae	*Aspidium* Hieron. (1896)	= *rivularioides*
arenicola	*Mesophlebion* Holtt. (1975)	*Mesophlebion* Holtt. (1975)
arenosa	*Thelypteris* A. R. Smith (1983)	= *Amauropelta*
arfakianum	*Polypodium* Bak. (1880)	*Sphaerostephanos* Holtt. (1982)
argentinum	*Aspidium* Hieron. (1896)	= *Amauropelta*
aridum	*Aspidium* D. Don (1825)	*Christella* Holtt. (1974)
arisanensis	*Dryopteris* Rosenst. (1915)	= *gracilescens* Bl.
aristata	*Goniopteris* Fée (1852)	= *Thelypteris* s.l.
aristeguietae	*Lastrea* Vareschi (1966)	= *Amauropelta*
armata	*Dryopteris* Rosenst. (1915)	= *imponens*
armata	*Lastrea* Copel. (1947)	= *Diplazium*
arthrothrix	*Polypodium* Hook. (1863)	= *Thelypteris* s.l.
arthrotricha	*Coryphopteris* Holtt. (1976)	*Coryphopteris* Holtt. (1976)
articulatum	*Nephrodium* Houlst. & Moore (1851)	*Pronephrium* Holtt. (1972)
artinexum	*Nephrodium* Clarke (1880)	= *clarkei* Bedd.
aspera	*Goniopteris* Presl (1836)	*Pronephrium* Holtt. (1972)
asperulum	*Aspidium* Fée (1866)	= *l'herminieri*
aspidioides	*Ceterach* Willd. (1810)	= *Amauropelta*
aspidioides	*Gymnogramme* Bl. (1828)	= *pozoi*
aspidioides	*Stegnogramma* Bl. (1828)	*Stegnogramma* Bl. (1828)
asplenioides	*Polypodium* Sw. (1810)	*Goniopteris* Presl (1836)
asplenioides	*Sphaerostephanos* J. Sm. (1839)	= *polycarpon* Bl.
assamica	*Dryopteris* Rosenst. (1917)	= *cylindrothrix*
assamicum	*Nephrodium* Bedd. (1893)	= *subelatum*
assurgens	*Dryopteris* Maxon (1944)	= *Amauropelta*

3

Epithet	Genus in which Originally Described	Present or Proposed Disposition
asterothrix	Goniopteris Fée (1852)	Goniopteris Fée (1852)
asymmetrica	Goniopteris Fée (1852)	= diversilobum
atasripii	Dryopteris Rosenst. (1917)	Sphaerostephanos Holtt. (1976)
athyriocarpa	Dryopteris Copel. (1908)	Coryphopteris Holtt. (1976)
athyrioides	Coryphopteris Holtt. (1976)	Coryphopteris Holtt. (1976)
atjehensis	Coryphopteris Holtt. (1976)	Coryphopteris Holtt. (1976)
atomiferum	Nephrodium Sod. (1883)	= cheilanthoides Kze
atrispora	Cyathea Domin (1930)	= dryopteroidea
var. varievestita	Dryopteris C. Chr. (1937)	= varievestita
atropurpurea	Dryopteris Hieron. (1907)	= Amauropelta
atrorubens	Aspidium Mett. (1869)	= Amauropelta
atrospinosa	Dryopteris Kjellb. & C. Chr. (1933)	Chingia Holtt. (1974)
atrovirens	Dryopteris C. Chr. (1907)	= Amauropelta
attenuata	Lastrea Brack. (1854)	Plesioneuron Holtt. (1975)
attenuatum	Aspidium Kunze ex Mett. (1858)	Amphineuron Holtt. (1977)
aubertii	Polypodium Desv. (1827)	Pseudophegopteris Holtt. (1969)
auctipinna	Pneumatopteris Holtt. (1973)	Pneumatopteris Holtt. (1973)
augescens	Aspidium Link (1841)	Christella Pic. Ser. (1977)
aureoglandulosus	Cyclosorus Ching & Shing (1983)	= Christella
aureoviridis	Dryopteris Rosenst. (1914)	= dayii
auriculata	Phegopteris J. Sm. (1875)	Cyclogramma Ching (1963)
auriculatum	Polypodium Wall. ex Hook. (1862)	= auriculata
auriculifera	Dryopteris v.A.v.R. (1922)	Mesophlebion Holtt. (1975)
aurita	Gymnogramme Hook. (1854)	Pseudophegopteris Ching (1963)
austera	Dryopteris Brause (1920)	Sphaerostephanos Holtt. (1982)
austrophilippina	Dryopteris Copel. (1929)	= sessilipinna
backeri	Dryopteris v.A.v.R. (1908)	= setigera
badia	Dryopteris v.A.v.R. (1914)	Coryphopteris Holtt. (1976)
bakeri	Nephrodium Harr. (1877)	Pronephrium Holtt. (1982)
balabacense	Pronephrium Zamora & Co (1980)	Pronephrium Zamora & Co (1980)
balansae	Cyclosorus Ching (1938)	Christella Holtt. (1976)
balansae	Polypodium Bak. (1891)	= torresianum
balbisii	Polypodium Spreng. (1821)	= Amauropelta
banaensis	Thelypteris Tard. & C. Chr. (1938)	= Macrothelypteris
bangii	Dryopteris C. Chr. (1907)	= Christella
baniensis	Dryopteris Rosenst. (1909)	= Amauropelta
baramensis	Dryopteris C. Chr. & Holtt. (1934)	Sphaerostephanos Holtt. (1982)
barbata	Goniopteris Fée (1852)	= ferox
bartlettii	Dryopteris Copel. (1929)	= glandulosum Bl.
basiattenuatum	Nephrodium Jenm. (1894)	= Amauropelta

Epithet	Genus in which Originally Described	Present or Proposed Disposition
basicurtata	*Pneumatopteris* Holtt. (1973)	*Pneumatopteris* Holtt. (1973)
basilaris	*Dryopteris* C. Chr. (1906)	= *productum* Kaulf.
basisora	*Dryopteris* Copel. (1911)	= *costulisora*
batacorum	*Dryopteris* Rosenst. (1914)	*Sphaerostephanos* Holtt. (1982)
var. *winkleri*	*Dryopteris* Rosenst. (1914)	= *truncatum* Poir.
batjanensis	*Dryopteris* Rosenst. (1917)	*Sphaerostephanos* Holtt. (1982)
batulantensis	*Sphaerostephanos* Holtt. (1982)	*Sphaerostephanos* Holtt. (1982)
beccarianum	*Nephrodium* Ces. (1876)	*Mesophlebion* Holtt. (1975)
beccarianum	*Meniscium* Ces. (1877)	*Pronephrium* Holtt. (1972)
beddomei	*Nephrodium* Bak. (1867)	*Parathelypteris* Ching (1963)
belensis	*Dryopteris* Copel. (1942)	*Plesioneuron* Holtt. (1975)
benguetense	*Athyrium* Christ (1907)	= *gracilescens* Bl.
benguetensis	*Cyclosorus* Copel. (1952)	= *parasiticum* L.
benoiteanum	*Polystichum* Gaud. (1827)	*Sphaerostephanos* Holtt. (1982)
berastagiensis	*Dryopteris* C. Chr. (1937)	= *inclusa*
bergianum	*Polypodium* Schlecht. (1825)	*Amauropelta* Holtt. (1974)
bermudianum	*Nephrodium* Bak. (1885)	= *Goniopteris*
berroi	*Dryopteris* C. Chr. (1912)	= *Christella*
berteroanum	*Aspidium* Fée (1866)	= *Amauropelta*
besukiensis	*Dryopteris* v.A.v.R. (1911)	= *immersum*
bewaniensis	*Chingia* Holtt. (1982)	*Chingia* Holtt. (1982)
biauritum	*Nephrodium* Bedd. (1892)	= *lebeufii*
bibrachiatum	*Nephrodium* Jenm. (1894)	= *Goniopteris*
bicolor	*Dryopteris* Bonap. (1924)	= *Thelypteris* s.l.
biformata	*Dryopteris* Rosenst. (1909)	= *Goniopteris*
biolleyi	*Aspidium* Christ (1901)	*Goniopteris* Pic. Ser. (1977)
bipinnata	*Dryopteris* Copel. (1911)	*Plesioneuron* Holtt. (1975)
blanda	*Phegopteris* Fée (1857)	= *Goniopteris*
blastophorus	*Cyclosorus* Alston (1956)	*Pneumatopteris* Holtt. (1974)
blepharis	*Thelypteris* A. R. Smith (1975)	= *Christella*
blumeana	*Grammitis* Presl (1836)	= *pozoi*
blumei	*Aspidium* Mett. (1856)	= *heterophyllum* Presl
blumei	*Thelypteris* Panigrahi (1975)	= *subpubescens* Bl.
boholensis	*Cyclosorus* Copel. (1952)	= *acrostichoides* Desv.
bojeri	*Polypodium* Hook. (1862)	= *cruciatum*
boliviensis	*Dryopteris* Morton (1938)	= *Amauropelta*
bonapartii	*Dryopteris* Rosenst. (1909)	= *Amauropelta*
boninensis	*Dryopteris* Koidz. (1924)	*Christella* Holtt. (1976)
boottii	*Aspidium* Tuerck. (1843)	= *Dryopteris*
boqueronensis	*Dryopteris* Hieron. (1907)	= *Amauropelta*
bordenii	*Dryopteris* Christ (1907)	= *heterocarpon*
borealis	*Coryphopteris* Holtt. (1976)	*Coryphopteris* Holtt. (1976)
boridensis	*Pneumatopteris* Holtt. (1982)	*Pneumatopteris* Holtt. (1982)
borneense	*Polypodium* Hook. (1863)	*Pronephrium* Holtt. (1982)
boydiae	*Aspidium* Eaton (1879)	*Christella* Holtt. (1976)
brachyodus	*Polypodium* Kze. (1834)	= *glandulosum* Desv. var.
brachypus	*Nephrodium* Sod. (1883)	= *Amauropelta*

5

Epithet	Genus in which Originally Described	Present or Proposed Disposition
brachypodum	Nephrodium Bak. (1886)	= Amauropelta
brackenridgei	Aspidium Mett. (1861)	= attenuata Brack.
brackenridgei	Polypodium Hook. (1863)	= glandulifera Brack.
bradei	Dryopteris Christ (1909)	= Amauropelta
braineoides	Polypodium Bak. (1888)	= erubescens Hook.
braithwaitei	Sphaerostephanos Holtt. (1977)	Sphaerostephanos Holtt. (1977)
brasiliensis	Dryopteris C. Chr. (1913)	= Steiropteris
brassii	Dryopteris C. Chr. (1937)	= beddomei
brauseanum	Pronephrium Holtt. (1972)	Pronephrium Holtt. (1972)
brausei	Dryopteris Hieron. (1907)	= Amauropelta
breutelii	Amauropelta Kze. (1843)	Amauropelta Kze. (1843)
brevipes	Pseudophegopteris Ching & Wu (1983)	Pseudophegopteris Ching & Wu (1983)
brevipilosa	Coryphopteris Holtt. (1976)	Coryphopteris Holtt. (1976)
brevipinna	Dryopteris C. Chr. (1906)	= exsculptum
brittonae	Dryopteris Maxon (1926)	Goniopteris Ching (1940)
brooksii	Dryopteris Copel. (1908)	Pneumatopteris Holtt. (1973)
bruneovillosa	Dryopteris C. Chr. (1906)	= polypodioides Hook.
brunneum	Polypodium C. Chr. (1906)	= paludosum Bl. & pyrrhorhachis
brunnescens	Dryopteris Kjellb. & C. Chr. (1933)	= appendiculata Bl.
bryanii	Dryopteris C. Chr. (1943)	Pneumatopteris Holtt. (1973)
buchtienii	Thelypteris A. R. Smith (1980)	= Steiropteris
bukoensis	Dryopteris Tagawa (1932)	Pseudophegopteris Holtt. (1969)
bulusanicum	Haplodictyum Holtt. (1973)	Pronephrium Holtt. (1982)
bungoensis	Dryopteris C. Chr. (1937)	= pterospora
bunnemeyeri	Thelypteris Reed (1968)	= gymnocarpa
burchardii	Dryopteris Rosenst. (1917)	= cuspidatum Bl.
burkartii	Goniopteris Abbiat. (1964)	Goniopteris Abbiat. (1964)
burkartii	Thelypteris Abbiat. (1964)	= Amauropelta
burmanicus	Cyclosorus Ching (1938)	Christella Holtt. (1976)
burundiensis	Christella Pic. Ser. (1983)	Christella Pic. Ser. (1983)
buwaldae	Pronephrium Holtt. (1972)	Christella Holtt. (1982)
cabrerae	Dryopteris Weathb. (1949)	= Thelypteris s.l.
caeca	Dryopteris Rosenst. (1909)	= Amauropelta
caespitosa	Phegopteris Fourn. (1872)	= blanda
var. ciliata	Lastrea Bedd.	= ciliata Benth.
calcaratum	Aspidium Bl. (1828)	Trigonospora Holtt. (1974)
var. B	Aspidium Thw. (1864)	= zeylanica Ching
calcareum	Nephrodium Jenm. (1886)	= venustum
calcicola	Dryopteris C. Chr. (1933)	= warburgii
callensii	Cyclosorus Alston (1956)	Christella Holtt. (1974)
callosum	Aspidium Bl. (1828)	Pneumatopteris Nakai (1933)
calva	Dryopteris Copel. (1910)	= gracilescens Bl.
calvata	Thelypteris Ching (1949)	= Parathelypteris
calvescens	Cyclosorus Ching (1938)	Christella Holtt. (1976)

6

Epithet	Genus in which Originally Described	Present or Proposed Disposition
calypso	*Thelypteris* L. D. Gomez (1976)	= *Goniopteris*
camarinensis	*Nannothelypteris* Holtt. (1976)	*Nannothelypteris* Holtt. (1976)
camerounensis	*Pseudocyclosorus* Holtt. (1974)	*Pseudocyclosorus* Holtt. (1974)
campii	*Thelypteris* A. R. Smith (1983)	= *Amauropelta*
camporum	*Polypodium* Lindm. (1903)	= *Amauropelta*
canum	*Nephrodium* Bak. (1867)	*Pseudocyclosorus* Holtt. & Grimes (1980)
canadasii	*Nephrodium* Sod. (1883)	= *Amauropelta*
canelensis	*Dryopteris* Rosenst. (1909)	= *Amauropelta*
canescens	*Polypodium* Bl. (1828)	Sphaerostephanos Holtt. (1982)
var. *degener*	*Dryopteris* Christ (1907)	= *clemensiae* var.
var. *lobata*	*Dryopteris* Christ (1907)	= *lobatus*
var. *novoguineensis*	*Dryopteris* Brause (1912)	= *brauseanum*
var. *subsimplicifolia*	*Dryopteris* Christ (1907)	= *melanophlebia*
forma *acrostichoides*	*Aspidium* Christ (1898)	= *celebicum* Bak.
forma *gymno- grammoides*	*Aspidium* Christ (1898)	= *batjanensis*
forma *nephrodii- formis*	*Aspidium* Christ (1898)	= *amboinense* Willd.
canlaonensis	*Dryopteris* Copel. (1929)	= *sessilipinna*
capitainei	*Aspidium* Fée & L'Herm. (1866)	= *l'herminieri*
caribaeum	*Nephrodium* Jenm. (1886)	= *resiniferum* Desv.
carolii	*Mesophlebion* Holtt. (1975)	*Mesophlebion* Holtt. (1975)
carolinensis	*Dryopteris* Hosok. (1936)	*Christella* Holtt. (1976)
carrii	*Sphaerostephanos* Holtt. (1982)	*Sphaerostephanos* Holtt. (1982)
cartilagidens	*Sphaerostephanos* Zamora & Co (1980)	*Sphaerostephanos* Zamora & Co (1980)
castanea	*Dryopteris* Tagawa (1935)	*Parathelypteris* Ching (1963)
catamarcensis	*Thelypteris* de la Sota (1973)	= *Amauropelta*
cataractorum	*Cyclosorus* Wagner & Greth. (1948)	Sphaerostephanos Holtt. (1982)
caucaense	*Nephrodium* Hieron. (1904)	= *Amauropelta*
caudata	*Leptogramma* Ching (1936)	= *tottoides*
caudatus	*Pseudocyclosorus* Holtt. (1965)	*Pneumatopteris* Holtt. (1973)
caudiculata	*Dryopteris* v.A.v.R. (1908)	= *microloncha*
caudiculata	*Dryopteris* Rosenst. (1911)	= *aquatilis*
caudiculata	*Lastrea* Presl (1851)	= *immersum*
caudiculatum	*Nephrodium* Presl (1851)	= *prismaticum*
caudipinna	*Thelypteris* Ching (1936)	*Trigonospora* Sledge (1981)
caulescens	*Sphaerostephanos* Holtt. (1982)	Sphaerostephanos Holtt. (1982)
cavaleriei	*Christella* Lev. (1915)	= *esquirolii*
cavitensis	*Lastrea* Copel. (1952)	= *costata*

Epithet	Genus in which Originally Described	Present or Proposed Disposition
celebica	*Leptogramma* Ching (1936)	*Stegnogramma* Holtt. (1982)
celebicum	*Acrostichum* Bak. (1901)	*Pronephrium* Holtt. (1972)
ceramica	*Phegopteris* v.A.v.R. (1908)	Amphineuron Holtt. (1977)
cesatiana	*Dryopteris* C. Chr. (1906)	= *beccarianum* (*Meniscium* Ces.)
chaerophylloides	*Polypodium* Poir. (1804)	= *Thelypteris* s.l.
chamaeotaria	*Dryopteris* Christ (1907)	= *granulosum*
changshaensis	*Macrothelypteris* Ching (1963)	*Macrothelypteris* Ching (1963)
chartacea	*Dryopteris* C. Chr. (1906)	= *grandis*
chaseana	*Thelypteris* Schelpe (1965)	= *Christella*
cheesmaniae	*Pneumatopteris* Holtt. (1973)	*Pneumatopteris* Holtt. (1973)
cheilanthoides	*Aspidium* Kze (1849)	*Amauropelta* Löve & Löve (1977)
cheilanthoides	*Polypodium* Bak. (1886)	= *polypodioides* Hook.
cheilocarpa	*Goniopteris* Fée (1852)	= *Cyclosorus*
cheilosora	*Gymnogramme* Fée (1857)	= *Amauropelta*
chimboracensis	*Thelypteris* A. R. Smith (1983)	= *stramineum* Sod.
chinensis	*Thelypteris* Ching (1936)	*Parathelypteris* Ching (1963)
chingii	*Cyclosorus* Liu (1983)	= ? *Pneumatopteris*
chlamydophora	*Dryopteris* Rosenst. (1928)	*Mesophlebion* Holtt. (1975)
chlorophylla	*Dryopteris* C. Chr. (1906)	= *lanosa*
christelloides	*Pneumatopteris* Holtt. (1973)	= *truncatum* Poir.
christensenii	*Dryopteris* Christ (1907)	= *Amauropelta*
christiana	*Dryopteris* Kodama (1924)	= *quelpaertensis*
christii	*Cyclosorus* Copel. (1960)	*Chingia* Holtt. (1974)
christii	*Dryopteris* C. Chr. (1906)	= *Meniscium*
christophersenii	*Dryopteris* C. Chr. (1943)	= *costata* Brack.
chrysodioides	*Meniscium* Fée (1852)	*Meniscium* Fée (1852)
chunii	*Thelypteris* Ching (1936)	*Cyclogramma* Tagawa (1938)
ciliata	*Lastrea* Hook. (1857)	*Trigonospora* Holtt. (1974)
ciliatum	*Aspidium* Benth. (1861)	*Trigonospora* Holtt. (1974)
cinereum	*Nephrodium* Sod. (1908)	= *Amauropelta*
clarkei	*Pleocnemia* Bedd. (1876)	*Christella* Holtt. (1974)
clarkei	*Polypodium* Bak. (1891)	= *mauiensis*
clavipilosa	*Chingia* Holtt. (1974)	*Chingia* Holtt. (1974)
clemensiae	*Dryopteris* Copel. (1931)	*Pronephrium* Holtt. (1972)
clypeolutatum	*Nephrodium* Desv. (1827)	= *Steiropteris*
coarctatum	*Aspidium* Kze (1845)	= *Amauropelta*
cochaensis	*Dryopteris* C. Chr. (1913)	= *Amauropelta*
columbiana	*Dryopteris* C. Chr. (1907)	= *Amauropelta*
comorensis	*Pneumatopteris* Holtt. (1973)	*Pneumatopteris* Holtt. (1973)
comosa	*Dryopteris* Morton (1938)	= *Steiropteris*
compacta	*Dryopteris* Copel. (1911)	= *hewittii*
concinnum	*Polypodium* Willd. (1810)	*Amauropelta* Pic. Ser. (1977)
conferta	*Dryopteris* Brause (1912)	*Sphaerostephanos* Holtt. (1976)
confluens	*Aspidium* Fée (1852)	= *Amauropelta*
confluens	*Pteris* Thunb. (1800)	*Thelypteris* Morton (1967)
conforme	*Nephrodium* Sod. (1883)	= *Amauropelta*

Epithet	Genus in which Originally Described	Present or Proposed Disposition
confusa	Dryopteris Copel. (1911)	= philippinum (Physematium Presl)
conioneuron	Nephrodium Fée (1852)	= opulentum
connectile	Polypodium Michx. (1803)	Phegopteris Watt (1870)
connexum	Nephrodium Kuhn (1870)	= elatum Boj.
consanguineum	Aspidium Fée (1866)	= Amauropelta
consanguineum	Polypodium Klotzsch (1847)	= Amauropelta
consanguineum	Polystichum Gaud. (1827)	= ? dentatum
consimilis	Gymnogramme Fée (1868)	= Amauropelta
consobrina	Dryopteris Maxon & Morton (1938)	= Meniscium
conspersoides	Aspidium Fée (1857)	= Thelypteris s.l.
conspersum	Nephrodium Schrad. (1824)	Christella Löve & Löve (1977)
conterminoides	Dryopteris C. Chr. (1906)	= fasciculatum
conterminum	Aspidium Willd. (1810)	= Amauropelta
contigua	Dryopteris Rosenst. (1917)	= hispidulum Dcne.
contingens	Macrothelypteris Ching (1963)	Macrothelypteris Ching (1963)
continuum	Aspidium Desv. (1811)	= interrupta Willd.
convergens	Sphaerostephanos Holtt. (1982)	Sphaerostephanos Holtt. (1982)
cooleyi	Thelypteris Proctor (1964)	= Amauropelta
copelandii	Thelypteris Reed (1968)	= majus
corazonense	Nephrodium Bak. (1877)	= Amauropelta
cordata	Phegopteris Fée (1852)	Goniopteris Brack. (1854)
cordifolia	Phegopteris v.A.v.R. (1913)	= menisciicarpon
coriacea	Dryopteris Brause (1920)	Coryphopteris Holtt. (1976)
cornuta	Dryopteris Maxon (1924)	= Amauropelta
correllii	Thelypteris A. R. Smith (1983)	= Amauropelta
cosmopolita	Lastrea Brack. (1872)	= torresianum
costale	Aspidium Mett. (1869)	= densisora
costale	Nephrodium Bak. (1874)	= leprieurii var.
costaricensis	Lastrea Copel. (1947)	= atrovirens
costata	Goniopteris Brack. (1854)	Pneumatopteris Holtt. (1973)
costatum	Nephrodium Bedd. (1867)	= penangianum
costulare	Nephrodium Bak. (1877)	= unita Kze
costulisora	Lastrea Copel. (1947)	Plesioneuron Holtt. (1975)
crassa	Dryopteris Copel. (1942)	Plesioneuron Holtt. (1975)
crassifolium	Aspidium Bl. (1828)	Mesophlebion Holtt. (1971)
crassinervia	Dryopteris C. Chr. (1934)	= tylodes
crassipes	Nephrodium Sod. (1893)	= Amauropelta
crassiuscula	Dryopteris C. Chr. & Maxon (1933)	= Amauropelta
creaghii	Nephrodium Bak. (1898)	= dayii
crenatum	Polypodium Sw. (1788)	= poiteana
crenulaeum	Nephrodium Jenm. (1896)	= Amauropelta
crenulatum	Pronephrium Holtt. (1972)	Pronephrium Holtt. (1972)
crespiana	Dryopteris Bosc. (1938)	= Amauropelta
cretacea	Thelypteris A. R. Smith (1971)	Christella Löve & Löve (1977)

9

Epithet	Genus in which Originally Described	Present or Proposed Disposition
crinipes	Nephrodium Hook. (1862)	Christella Holtt. (1974)
cristatum	Polypodium L. (1753)	= Dryopteris
croftii	Plesioneuron Holtt. (1981)	Plesioneuron Holtt. (1981)
crossii	Polypodium Bak. (1891)	= Amauropelta
cruciatum	Aspidium Willd. (1810)	Pseudophegopteris Holtt. (1969)
cryptum	Polypodium Underw. & Maxon (1902)	Goniopteris Ching (1940)
ctenoides	Polypodium Jenm. (1897)	= rude Kze
ctenolobum	Plesioneuron Holtt. (1975)	Plesioneuron Holtt. (1975)
cubanum	Polypodium Bak. (1867)	= cordata
cucullatum	Aspidium Bl. (1828)	= unitum (Polypodium L.)
var. mucronata	Dryopteris Christ (1907)	= unitum (Polypodium L.) var.
cumingianum	Aspidium Kze (1840)	Goniopteris de Joncheere & Sen (1981)
cumingianum	Anisocampium Presl (1849)	= aristata
cumingii	Meniscium Fée (1852)	= triphyllum
cuneata	Dryopteris C. Chr. (1913)	Goniopteris Brade (1972)
cunninghamii	Phegopteris Mett. (1856)	= pennigerum Forst.
curta	Dryopteris Christ (1907)	= Goniopteris
cuspidatum	Polypodium Roxb. (1844)	= repanda
cuspidatum	Meniscium Bl. (1828)	Pronephrium Holtt. (1972)
var. longifrons	Meniscium Clarke (1880)	= lakhimpurensis
cutiataensis	Dryopteris Brade (1951)	Goniopteris Brade (1972)
cyatheoides	Aspidium Kaulf. (1824)	Christella Holtt. (1976)
var. cyatheoides	Dryopteris C. Chr. (1906)	= boydiae
var. depauperata	Aspidium Hillebr. (1888)	= boydiae
cyclocarpa	Pseudophegopteris Holtt. (1965)	Pseudophegopteris Holtt. (1965)
cyclolepis	Athyrium C. Chr. & Tard. (1934)	= confluens Thunb.
cylindrothrix	Dryopteris Rosenst. (1913)	Christella Holtt. (1974)
cyrtocaulos	Dryopteris v.A.v.R. (1922)	Sphaerostephanos Holtt. (1982)
cyrtomioides	Dryopteris C. Chr. (1924)	Stegnogramma Ching (1936)
cystodioides	Plesioneuron Holtt. (1982)	Plesioneuron Holtt. (1982)
cystopteroides	Athyrium Eaton (1858)	Parathelypteris Ching (1963)
dalhousiana	Goniopteris Fée (1857)	= repanda
dasyphylla	Dryopteris C. Chr. (1906)	= villosa
dasypoda	Thelypteris Morton (1973)	= aspidioides Bl.
dayii	Aspidium Bedd. (1888)	Metathelypteris Holtt. (1974)
daymanianus	Sphaerostephanos Holtt. (1982)	Sphaerostephanos Holtt. (1982)
debile	Nephrodium Bak. (1880)	Pronephrium Holtt. (1972)
debilis	Phegopteris Mett. (1864)	Sphaerostephanos Holtt. (1982)
decadens	Nephrodium Bak. (1886)	Sphaerostephanos Holtt. (1977)
decipiens	Nephrodium Clarke (1880)	Metathelypteris Ching (1963)
decora	Dryopteris Domin (1913)	= terminans

Epithet	Genus in which Originally Described	Present or Proposed Disposition
decrescens	Aspidium Mett. (1858)	= Amauropelta
decrescens	Thelypteris Proctor (1981)	= Amauropelta
decumbens	Dryopteris C. Chr. (1906)	= l'herminieri
decurrens	Lastrea J. Sm. (1846)	= decursivepinnatum
decurrentialata	Gymnogramme Hook. (1864)	= Diplazium
decursivepinnatum	Polypodium van Hall (1836)	Phegopteris Fée (1852)
decurtatum	Asplenium Link (1841)	= Amauropelta
decurtatum	Aspidium Kze (1850)	= truncatum Poir.
decussatum	Polypodium L. (1753)	= Steiropteris
deficiens	Pneumatopteris Holtt. (1973)	Pneumatopteris Holtt. (1973)
deflectens	Dryopteris C. Chr. (1937)	= Amauropelta
deflexum	Nephrodium Presl (1825)	Amauropelta Löve & Löve (1977)
degener	Dryopteris Christ (1907)	= clemensiae
degeneri	Cyclosorus Copel. (1949)	= decadens
dejectum	Nephrodium Jenm. (1895)	= Thelypteris s.l.
delicatula	Phegopteris Fée (1866)	= Amauropelta
deltiptera	Dryopteris Copel. (1942)	= superba
deltoideum	Polypodium Sw. (1788)	Steiropteris Pic. Ser. (1973)
demeraranum	Polypodium Bak. (1866)	= Amauropelta
deminuens	Cyclosorus Holtt. (1965)	= atasripii
densa	Dryopteris Maxon (1944)	= Amauropelta
densiloba	Dryopteris C. Chr. (1906)	= Steiropteris
densisora	Dryopteris C. Chr. (1906)	Steiropteris Pic. Ser. (1973)
dentatum	Polypodium Forssk. (1773)	Christella Brownsey & Jermy (1973)
desvauxii	Aspidium Mett. (1868)	= Meniscium
deversum	Aspidium Kze (1850)	= patens Sw.
devolvens	Nephrodium Bak. (1885)	= Goniopteris
dewevrei	Dryopteris Bonap. (1924)	= afra
dianae	Polypodium Hook. (1862)	Pseudophegopteris Holtt. (1969)
diaphana	Dryopteris Brause (1920)	Coryphopteris Holtt. (1976)
dicarpum	Nephrodium Fée (1852)	= ? hispidulum Dcne
dichotomum	Polypodium Panz. (1876)	acuminatum Houtt.
dichrotricha	Dryopteris Copel. (1911)	Sphaerostephanos Holtt. (1982)
dichrotricha	Dryopteris Copel. (1912)	= dichrotrichoides
dichrotrichoides	Dryopteris v.A.v.R (1917)	Sphaerostephanos Holtt. (1975)
dicksonioides	Phegopteris Kuhn (1864)	= Ctenitis glabra
dicranogramma	Dryopteris v.A.v.R. (1922)	Pneumatopteris Holtt. (1973)
dictyoclinoides	Stegnogramma Ching (1936)	Stegnogramma Ching (1936)
didymochlaenoides	Nephrodium Clarke (1880)	Coryphopteris Holtt. (1974)
didymosorum	Nephrodium Parish ex Bedd. (1866)	= parasiticum L.
dilatatum	Polypodium Hoffm. (1795)	= Dryopteris
dimidiolobata	Sphaerostephanos Holtt. (1982)	Sphaerostephanos Holtt. (1982)
diminuens	Cyclosorus Holtt. (1965)	= atasripii
diminuta	Dryopteris Copel. (1929)	Pronephrium Holtt. (1972)

11

Epithet	Genus in which Originally Described	Present or Proposed Disposition
dimorpha	*Dryopteris* Brause (1920)	*Sphaerostephanos* Holtt. (1982)
dimorphus	*Cyclosorus* Copel. (1954)	= *majus*
diplazioides	*Aspidium* Mett. (1858)	= *Amauropelta*
diplazioides	*Gymnogramme* Desv. (1827)	*Amauropelta* Löve & Löve (1977)
dispar	*Dryopteris* Maxon & Morton (1938)	= *Meniscium*
dissimile	*Nephrodium* Schrad. (1824)	= *patens* Sw.
dissimulans	*Dryopteris* C. Chr. (1913)	= *Goniopteris*
dissitifolia	*Stegnogramma* Holtt. (1982)	*Stegnogramma* Holtt. (1982)
distans	*Nephrodium* Hook (1862)	*Christella* Holtt. (1974)
distans	*Polypodium* D. Don (1825)	= *pyrrhorhachis*
distans	*Polypodium* Racib. (1898)	= *paludosum* Bl.
var. *adnatum*	*Polypodium* Clarke (1880)	= *pyrrhorhachis*
var. *glabratum*	*Polypodium* Clarke (1880)	= *pyrrhorhachis* var.
var. *minor*	*Polypodium* Clarke (1880)	= *rectangulare*
distincta	*Dryopteris* Copel. (1942)	*Amphineuron* Holtt. (1977)
divergens	*Dryopteris* Rosenst. (1914)	= *crassifolium* Bl.
diversifolia	*Dryopteris* v.A.v.R. (1908)	= *immersum*
diversilobum	*Nephrodium* Presl (1851)	*Sphaerostephanos* Holtt. (1975)
subvar. *lanceola*	*Dryopteris* Christ (1907)	= *rhombea*
subvar. *rhombea*	*Dryopteris* Christ (1907)	= *rhombea*
diversisora	*Dryopteris* Copel. (1938)	*Coryphopteris* Holtt. (1976)
diversivenosa	*Dryopteris* v.A.v.R. (1918)	= *dayii*
doctersii	*Plesioneuron* Holtt. (1975)	*Plesioneuron* Holtt. (1975)
dodsonii	*Thelypteris* A. R. Smith (1983)	= *Amauropelta*
domingense	*Polypodium* Spr. (1827)	*Goniopteris* Fée (1866)
dominicensis	*Dryopteris* C. Chr. (1909)	= *Amauropelta*
donnell-smithii	*Dryopteris* Maxon (1909) (nom. nud.)	= *paucipinnatum*
doodioides	*Dryopteris* Copel. (1936)	*Sphaerostephanos* Holtt. (1977)
dryas	*Plesioneuron* Holtt. (1975)	*Plesioneuron* Holtt. (1975)
dryopteroidea	*Alsophila* Brause (1920)	*Plesioneuron* Holtt. (1975)
dubreuillianum	*Polystichum* Gaud. (1827)	= *cyatheoides*
duchassaigniana	*Phegopteris* Fée (1866)	= *Amauropelta*
duclouxii	*Dryopteris* Christ (1907)	= *esquirolii*
dulitense	*Mesophlebion* Holtt. (1975)	*Mesophlebion* Holtt. (1975)
dumetorum	*Dryopteris* Maxon (1944)	= *Amauropelta*
duplosetosus	*Cyclosorus* Copel. (1952)	= *lastreoides*
dura	*Dryopteris* Copel. (1910)	*Coryphopteris* Holtt. (1976)
dutrai	*Dryopteris* Dutra (1940)	= *Amauropelta*
eberhardtii	*Dryopteris* Christ (1908)	= *esquirolii*
eburnea	*Pneumatopteris* Holtt. (1982)	*Pneumatopteris* Holtt. (1982)
eburneum	*Aspidium* Wall. (1828) (nom. nud.)	= *canum*
ecallosus	*Cyclosorus* Holtt. (1947)	*Pneumatopteris* Holtt. (1973)
echinatum	*Aspidium* Mett. (1864)	*Mesophlebion* Holtt. (1975)

Epithet	Genus in which Originally Described	Present or Proposed Disposition
echinospora	Dryopteris v.A.v.R. (1920)	Sphaerostephanos Holtt. (1982)
ecklonii	Aspidium Kze. (1836)	= interrupta Willd.
edanyoi	Cyclosorus Copel. (1952)	Pronephrium Holtt. (1982)
efogensis	Sphaerostephanos Holtt. (1982)	Sphaerostephanos Holtt. (1982)
egenolfioides	Pneumatopteris Holtt. (1973)	Pneumatopteris Holtt. (1973)
eggersii	Nephrodium Hieron. (1904)	Goniopteris Alston (1957)
ekmanii	Thelypteris A. R. Smith ex Lellinger (1984)	= Thelypteris sens. lat.
ekutiensis	Sphaerostephanos Holtt. (1982)	Sphaerostephanos Holtt. (1982)
elatum	Aspidium Boj. (1837)	Sphaerostephanos Holtt. (1974)
elatum	Aspidium Kuhn (1868)	= venulosum Hook.
elatior	Aspidium Fée (1869)	= Amauropelta
elegans	Ampelopteris Kze (1848)	= prolifera
elegans	Dryopteris Koidz. (1924)	= torresianum var. calvata
var. subtripinnata	Dryopteris Tagawa (1933)	= viridifrons
elegantulum	Nephrodium Sod. (1893)	= Amauropelta
eliasii	Polypodium Sennen & Pau (1911)	= pozoi
elliptica	Dryopteris Rosenst. (1917)	Sphaerostephanos Holtt. (1975)
ellipticus	Cyclosorus Copel. (1947)	= norrisii & elliptica Rosenst.
elmerorum	Dryopteris Copel. (1929)	= norrisii
elongata	Phegopteris Fourn. (1872)	= concinnum var.
elwesii	Nephrodium Bak. (1874)	Oreopteris Holtt. (1974)
eminens	Nephrodium Bak. (1880)	Sphaerostephanos Holtt. (1982)
endertii	Dryopteris C. Chr. (1937)	Mesophlebion Holtt. (1975)
engelii	Dryopteris Hieron. (1907)	= Amauropelta
engleriana	Dryopteris Brause (1912)	Coryphopteris Holtt. (1976)
var. hirta	Dryopteris C. Chr. (1937)	= fasciculatum
ensifera	Dryopteris Tagawa (1937)	Christella Holtt. (1975)
ensiformis	Dryopteris C. Chr. (1913)	Meniscium Pic. Ser. (1968)
ensipinna	Dryopteris Brause (1920)	= savaiense
epaleata	Dryopteris C. Chr. (1934)	= richardsii
equitans	Nephrodium Christ (1906)	= Goniopteris
erectus	Cyclosorus Copel. (1952)	Sphaerostephanos Holtt. (1975)
eriocarpa	Glaphyropteridopsis Ching (1963)	Glaphyropteridopsis Ching (1963)
eriosorus	Aspidium Fée (1873)	= Amauropelta
erubescens	Polypodium Hook. (1862)	= Glaphyropteridopsis Ching (1963)
var. amboinense	Polypodium Bak. (1867)	= ceramica
erwinii	Sphaerostephanos Holtt. (1982)	Sphaerostephanos Holtt. (1982)
espinosae	Dryopteris Hicken (1913)	= Dryopteris

Epithet	Genus in which Originally Described	Present or Proposed Disposition
esquirolii	*Dryopteris* Christ (1907)	*Pseudocyclosorus* Ching (1963)
etchichuryi	*Nephrodium* Hicken (1907)	= *Goniopteris*
euaensis	*Dryopteris* Copel. (1931)	= *harveyi*
euchlorum	*Polypodium* Sod. (1893)	= *Amauropelta*
eugracilis	*Lastrea* Copel. (1947)	= *beddomei* var.
euphlebius	*Cyclosorus* Ching (1938)	*Christella* Holtt. (1976)
eurostotrichum	*Nephrodium* Bak. (1880)	= *distans* Hook.
euryphylla	*Dryopteris* Rosenst. (1917)	*Pronephrium* Holtt. (1972)
eusorum	*Aspidium* Thw. (1864)	= *truncatum* Poir.
euthythrix	*Thelypteris* A. R. Smith (1983)	= *Amauropelta*
evolutum	*Nephrodium* Clarke & Bak. (1892)	*Christella* Holtt. (1974)
var. *B*	*Nephrodium* Bedd. (1892)	= *gustavii*
excellens	*Aspidium* Bl. (1828)	= *penniger um* var. *excellens*
excelsum	*Polypodium* Bak. (1874)	= *glandulifera* Brack.
excisus	*Pseudocyclosorus* Holtt. (1965)	*Pneumatopteris* Holtt. (1973)
excrescens	*Dryopteris* Copel. (1929)	= *glandulosum* Bl.
exiguum	*Aspidium* Mett. (1858)	= *nervosa* Fée, *inaequilobata* & *philippinum* (*Physematium* Presl)
exindusiatus	*Sphaerostephanos* Holtt. (1982)	*Sphaerostephanos* Holtt. (1982)
expansa	*Gymnogramme* Fée (1869)	= *Amauropelta*
exsculptum	*Acrostichum* Bak. (1888)	*Pronephrium* Holtt. (1972)
exsudans	*Aspidium* Fourn. (1872)	= *Amauropelta*
extensum	*Aspidium* Bl. (1828)	= *opulentum*
var. *laterepens*	*Nephrodium* Clarke (1880)	= *appendiculatum* Presl
var. *microsorum*	*Nephrodium* Clarke (1880)	= *appendiculatum* Presl
var. *minor*	*Nephrodium* Bedd. (1865)	= *zeylanica* Ching
exuta	*Thelypteris* A. R. Smith (1983)	= *Amauropelta*
fadenii	*Thelypteris* Fosberg & Sachet (1972)	= *parasiticum* L.
fairbankii	*Lastrea* Bedd. (1867)	= *confluens* Thunb.
falcatilobum	*Mesophleblon* Holtt. (1982)	*Mesophlebion* Holtt. (1982)
falcatum	*Meniscium* Liebm. (1849)	*Meniscium* Liebm. (1849)
falcatipinnula	*Dryopteris* Copel. (1911)	*Plesioneuron* Holtt. (1975)
falcatula	*Dryopteris* Christ (1907)	= *hispidulum* Dcne
falciloba	*Lastrea* Hook. (1857)	*Pseudocyclosorus* Ching (1963)
var. *B*	*Nephrodium* Hook. (1862)	= *zeylanica* Ching
farinosa	*Dryopteris* Brause (1920)	= *munda*
fasciculatum	*Aspidium* Fourn. (1873)	*Coryphopteris* Holtt. (1976)
fatuhivensis	*Dryopteris* E. Brown (1931)	= *Dryopteris*
feei	*Thelypteris* Moxley (1921)	= *puberulum*
fendleri	*Acrostichum* Bak. (1887)	= *guyanensis*
fendleri	*Aspidium* Eaton (1860)	*Steiropteris* Pic. Ser. (1973)

14

Epithet	Genus in which Originally Described	Present or Proposed Disposition
fenixii	Sphaerostephanos Holtt. (1975)	= irayensis
ferox	Aspidium Bl. (1828)	Chingia Holtt. (1971)
var. calvescens	Dryopteris Christ (1907)	= christii Copel.
finisterrae	Dryopteris Brause (1912)	Pneumatopteris Holtt. (1982)
firmum	Nephrodium Jenm. (1879)	= Amauropelta
firmulum	Polypodium Bak. (1893)	Pronephrium Holtt. (1972)
fischeri	Aspidium Mett. (1858)	= Amauropelta
flaccidum	Aspidium Bl. (1828)	Metathelypteris Ching (1963)
flaccidus	Cyclosorus Ching & Liu (1983)	= ? Pneumatopteris
flavovirens	Dryopteris Rosenst. (1912)	= dayii
flavoviridis	Sphaerostephanos Holtt. (1982)	Sphaerostephanos Holtt. (1982)
flexile	Aspidium Christ (1902)	= Cyclogramma
fluminalis	Thelypteris A. R. Smith (1983)	= Amauropelta
foliolosus	Sphaerostephanos Holtt. (1982)	Sphaerostephanos Holtt. (1982)
formosa	Dryopteris Nakai (1931)	= japonicum var.
forsteri	Thelypteris Morton (1967)	= invisum Forst.
foxii	Dryopteris Christ (1907)	= ligulata
foxworthyi	Sphaerostephanos Holtt. (1982)	Sphaerostephanos Holtt. (1982)
fragile	Polypodium Bak. (1877)	Metathelypteris Holtt. (1974)
fragilis	Alsophila Zoll. & Moritzi (1844)	= lineatum Bl.
fragans	Polypodium L. (1753)	= Dryopteris
francii	Dryopteris Copel. (1929)	= epaleata
francoanum	Aspidium Fourn. (1872)	Goniopteris Löve & Löve (1977)
fraseri	Aspidium Mett. ex Kuhn (1869)	= Goniopteris
friesii	Dryopteris Brause (1914)	Christella Holtt. (1974)
frigida	Aspidium Christ (1906)	= Amauropelta
fuchsii	Plesioneuron Holtt. (1975)	Plesioneuron Holtt. (1975)
fuertesii	Dryopteris Brause (1913)	= Goniopteris
fukiensis	Cyclosorus Ching (1938)	Christella Holtt. (1976)
fulgens	Dryopteris Brause (1920)	Plesioneuron Holtt. (1975)
funckii	Aspidium Mett. (1864)	= Amauropelta
furva	Dryopteris Maxon (1944)	= Amauropelta
galanderi	Aspidium Hieron. (1897)	= Amauropelta
gamblei	Pseudocyclosorus Holtt & Grimes (1980)	Pseudocyclosorus Holtt & Grimes (1980)
gardneri	Pronephrium Holtt. (1971)	Pronephrium Holtt. (1971)
gardneriana	Nephrodium Bak. (1870)	= densiloba
gaudichaudii	Sphaerostephanos Holtt. (1979)	Sphaerostephanos Holtt. (1979)
gemmulifera	Dryopteris Hieron. (1907)	= Goniopteris
germaniana	Phegopteris Fée (1866)	= Amauropelta
germanii	Aspidium l'Hermin. (1866)	= hispidulum Dcne

15

Epithet	Genus in which Originally Described	Present or Proposed Disposition
geropogon	*Aspidium* Fée (1865)	= *albicaule*
ghiesbreghtii	*Polypodium* Hook. (1864)	*Goniopteris* J. Sm. (1872)
gigantea	*Meniscium* Mett. (1856)	= *simplicifrons*
gigantea	*Cheilanthes* Ces. (1877)	= *polypodioides* Hook.
giluwense	*Pronephrium* Holtt. (1982)	*Pronephrium* Holtt. (1982)
glabellus	*Cyclosorus* Ching (ined)	= ? *Pneumatopteris*
glaber	*Cyclosorus* Copel. (1952)	*Pneumatopteris* Holtt. (1973)
glaberrima	*Aspidium* Richard (1834)	*Pneumatopteris* Holtt. (1973)
glabratum	*Aspidium* Kuhn (1868)	= *Lunathyrium*
var. *hirsuta*	*Thelypteris* Tard. (1952)	= *cruciatum*
gladiata	*Dryopteris* C. Chr. (1916)	= *unita* Kze
glandafra	*Thelypteris* Viane (1985)	= *Pneumatopteris*
glandulifera	*Goniopteris* Brack. (1854)	*Pneumatopteris* Holtt. (1973)
glanduliferum	*Aspidium* Karst. (1847)	= *Amauropelta*
glanduligerum	*Aspidium* Kze (1837)	*Parathelypteris* Ching (1963)
glandulosa	*Trigonospora* Sledge (1981)	*Trigonospora* Sledge (1981)
glandulosum	*Aspidium* Bl. (1828)	*Pronephrium* Holtt. (1972)
var. *laetestrigosum*	*Nephrodium* Clarke (1880)	= *articulatum*
glandulosum	*Polypodium* Desv. (1811)	*Steiropteris* Pic. Ser. (1973)
glandulosolanosa	*Dryopteris* C. Chr. (1937)	= *Amauropelta*
glaucescens	*Dryopteris* Brause (1920)	= *novoguineensis*
glaucostipes	*Nephrodium* Bedd. (1892)	= *latebrosum*
glaziovii	*Aspidium* Christ (1902)	= *Amauropelta*
gleichenioides	*Aspidium* Christ (1904)	= *pterifolium*
globulifera	*Lastrea* Brack. (1854)	*Amauropelta* Holtt. (1977)
glochidiata	*Dryopteris* C. Chr. (1913)	*Goniopteris* Brade (1972)
glutinosa	*Dryopteris* C. Chr. (1937)	= *Amauropelta*
goedenii	*Dryopteris* Rosenst. (1907)	= *Christella*
goggilodus	*Aspidium* Schkuhr (1809)	= *gongylodes* Schkuhr
goldianum	*Aspidium* Goldie (1822)	= *Dryopteris*
goldii	*Dryopteris* C. Chr. (1913)	*Goniopteris* Brade (1972)
gongylodes	*Aspidium* Schkuhr (1809)	*Cyclosorus* Link (1833)
gracilenta	*Polypodium* Jenm. (1897)	= *Amauropelta*
gracilescens	*Aspidium* Bl. (1828)	*Metathelypteris* Ching (1963)
gracilescens	*Lastrea* Bedd. (1883)	= *beddomei, glanduligerum & hirsutipes*
var. *chinensis*	*Dryopteris* Christ (1909)	= *hirsutipes*
gracilis	*Dryopteris* Copel. (1929)	= *beddomei*
grammica	*Polypodium* Spr. (1822)	= *decussatum*
grammitoides	*Aspidium* Christ (1898)	*Parathelypteris* Ching (1963)
grandescens	*Sphaerostephanos* Holtt. (1982)	*Sphaerostephanos* Holtt. (1982)
grandis	*Thelypteris* A. R. Smith (1971)	= *Christella*
grantii	*Dryopteris* Copel. (1932)	*Amauropelta* Holtt. (1977)
granulosum	*Polypodium* Presl (1825)	*Pronephrium* Holtt. (1972)
grayii	*Nephrodium* Jenm. (1908)	= *glandulosum* Desv.
gregarius	*Cyclosorus* Copel. (1943)	*Sphaerostephanos* Holtt. (1982)
grenadense	*Nephrodium* Jenm. (1894)	= *l'herminieri*
gretheri	*Lastrea* Wagner (1948)	*Christella* Holtt. (1977)

Epithet	Genus in which Originally Described	Present or Proposed Disposition
griffithianum	Nephrodium Fée (1852)	= ? Christella
griffithii	Dictyocline Moore (1855)	Stegnogramma Iwats. (1963)
griffithii	Polypodium Hook. (1862)	= pyrrhorhachis
griseum	Nephrodium Bak. (1867)	Pseudocyclosorus Holtt. & Grimes (1980)
grossa	Lastrea Presl (1849)	= Amauropelta
grunowii	Aspidium Bolle (1855)	= striatum
guadalupense	Polypodium Wikstr. (1826)	= domingensis
guadalupensis	Goniopteris Fée (1866)	Goniopteris Fée (1866)
guamensis	Christella Holtt. (1977)	Christella Holtt. (1977)
gueintzianum	Aspidium Mett. (1858)	Christella Holtt. (1974)
guineensis	Dryopteris Christ (1909)	Christella Holtt. (1974)
gustavii	Nephrodium Bedd. (1893)	Christella Holtt. (1974)
guyanense	Meniscium Fée (1852)	Meniscium Fée (1852)
gymnocarpa	Dryopteris Copel. (1910)	Stegnogramma Iwats. (1963)
gymnopodum	Nephrodium Bak. (1894)	Coryphopteris Holtt. (1976)
gymnopteridifrons	Dryopteris Hayata (1919)	Pronephrium Holtt. (1972)
gymnorachis	Sphaerostephanos Holtt. (1982)	Sphaerostephanos Holtt. (1982)
habbemensis	Dryopteris Copel. (1942)	Coryphopteris Holtt. (1976)
haenkeanum	Nephrodium Presl (1849)	= unitum (Polypodium L.)
hainanensis	Cyclosorus Ching (1964)	= parasiticum L.
hakgalensis	Amauropelta Holtt. (1981)	Amauropelta Holtt. (1981)
halconensis	Cyclosorus Copel. (1952)	= latebrosum
hallieri	Aspidium Christ (1905)	Mesophlebion Holtt. (1975)
hamifera	Dryopteris v.A.v.R. (1914)	Sphaerostephanos Holtt. (1982)
handroi	Dryopteris Brade (1965)	Meniscium Brade (1972)
harcourtii	Dryopteris Domin (1929)	= Amauropelta
harrisii	Thelypteris Proctor (1981)	= Amauropelta
harrisonii	Nephrodium Bak. (1891)	= francoanum
harveyi	Aspidium Mett. (1869)	Christella Holtt. (1976)
hassleri	Dryopteris Christ (1907)	= Amauropelta
hastata	Goniopteris Fée (1866)	Goniopteris Fée (1866)
hastatopinnata	Dryopteris Brause (1920)	Sphaerostephanos Holtt. (1982)
hastiloba	Dryopteris C. Chr. (1937)	= Amauropelta
hatchii	Thelypteris A. R. Smith (1973)	Goniopteris Löve & Löve (1977)
hatschbachii	Thelypteris A. R. Smith (1980)	= Steiropteris
hattorii	Dryopteris H. Ito (1935)	Metathelypteris Ching (1963)
hawaiiensis	Thelypteris Reed (1968)	= sandwicensis
heimeri	Dryopteris C. Chr. (1909)	= Amauropelta
helliana	Phegopteris Fée (1852)	= aubertii
hellwigensis	Sphaerostephanos Holtt. (1982)	Sphaerostephanos Holtt. (1982)
helveolum	Aspidium Fée (1869)	= Amauropelta
hemitelioides	Dryopteris Christ (1909)	= striatum
hendersonii	Sphaerostephanos Holtt. (1982)	Sphaerostephanos Holtt. (1982)

Epithet	Genus in which Originally Described	Present or Proposed Disposition
henriquesii	Polypodium Bak. (1887)	Pseudophegopteris Holtt. (1969)
herbacea	Thelypteris Holtt. (1947)	= hirsutipes
hernaezii	Sphaerostephanos Holtt. (1982)	Sphaerostephanos Holtt. (1982)
herusiana	Aspidium Fourn. (1873)	= richardsii
herzogii	Dryopteris Rosenst. (1913)	= Amauropelta
heterocarpon	Aspidium Bl. (1828)	Sphaerostephanos Holtt. (1974)
heteroclitum	Polypodium Desv. (1811)	Amauropelta Pic. Ser. (1977)
heterodon	Aspidium Bl. (1828)	= polycarpon Bl.
heterophyllum	Haplodictyum Presl (1851)	Pronephrium Holtt. (1982)
heteropterum	Nephrodium Desv. (1827)	Amauropelta Holtt. (1974)
heterotricha	Dryopteris C. Chr. (1913)	= Goniopteris
hewittii	Dryopteris Copel. (1909)	Pronephrium Holtt. (1982)
hexagonopterum	Polypodium Michx. (1803)	Phegopteris Fée (1852)
hickenii	Cyclosorus Abbiatti (1965)	= conspersum
hieronymusii	Dryopteris C. Chr. (1907)	= Amauropelta
hildae	Thelypteris Proctor (1985)	= Goniopteris
hillii	Polypodium Bak. (1874)	= ghiesbrechtii
hilsenbergii	Nephrodium Presl (1851)	Christella Holtt. (1976)
himalaica	Leptogramma Ching (1936)	Stegnogramma Iwats. (1963)
himalayensis	Dryopteris C. Chr. (1934)	= auriculata
hirsutipes	Nephrodium Clarke (1880)	Coryphopteris Holtt. (1974)
hirsutisquama	Dryopteris Hayata (1915)	= uraiensis
hirsutum	Aspidium Kunze ex Mett. (1858)	Sphaerostephanos Holtt. (1975)
hirsutum	Aspidium sensu Christ in Ann. Jard. Bot. Buit. 15 (1898)	= polycarpon Bl.
hirticarpa	Dryopteris Ching (1931)	= hirtisora
hirtirachis	Dryopteris C. Chr. (1917)	Pseudophegopteris Holtt. (1969)
hirtisora	Dryopteris C. Chr. (1926)	Sphaerostephanos Holtt. (1979)
hirtisquamata	Dryopteris Hayata (1915)	= uraiensis
hirtopilosa	Dryopteris Rosenst. (1917)	= hispidulum Dcne
hispida	Dryopteris Brause (1920)	= hispiduliformis
hispidifolia	Dryopteris v.A.v.R. (1915)	Sphaerostephanos Holtt. (1982)
hispidula	Dryopteris sensu C. Chr., I.F. (1906)	= smithianum
hispiduliformis	Dryopteris C. Chr. (1934)	Sphaerostephanos Holtt. (1982)
hispidulum var. solutum	Aspidium Dcne (1834) Aspidium Miq. (1868)	Christella Holtt. (1976) = solutus
hispidulum	Nephrodium Christ (1906)	= hispidifolia
hispidulum	Nephrodium Peter (1929)	= dentatum
hoalensis	Sphaerostephanos Holtt. (1982)	Sphaerostephanos Holtt. (1982)
hochreutinieri	Dryopteris Christ (1912)	Sphaerostephanos Holtt. (1977)

18

Epithet	Genus in which Originally Described	Present or Proposed Disposition
hokouensis	Cyclosorus Ching (1949)	Christella Holtt. (1976)
holmei	Nephrodium Bak. (1891)	= clypeolutatum var.
holodictya	Thelypteris Kramer (1969)	= Meniscium
holophyllum	Polypodium Bak. (1888)	= menisciicarpon
hondurensis	Thelypteris L. Gómez (1982)	= Goniopteris
hookeri	Aspidium Bak. (1867)	= opulentum
hookeri	Nephrodium Houtt. & Moore (1851)	= arbuscula Willd.
hopeanum	Nephrodium Bak. (1874)	Plesioneuron Holtt. (1975)
hopei	Nephrodium Bak. (1891)	= appendiculata Bl.
horizontale	Athyrium Rosenst. (1912)	Coryphopteris Holtt. (1976)
horrens	Dryopteris Hieron. (1907)	= Amauropelta
horridipes	Dryopteris v.A.v.R. (1918)	Chingia Holtt. (1974)
hosei	Meniscium Bak. (1886)	Pronephrium Holtt. (1972)
var. sumbensis	Meniscium v.A.v.R. (1908)	= celebicum Bak.?
hostmannii	Polypodium Klotzsch (1847)	= Meniscium
hottensis	Dryopteris C. Chr. (1937)	= Steiropteris
houii	Cyclosorus Ching (1949)	= Christella
hubrechtensis	Coryphopteris Holtt. (1976)	Coryphopteris Holtt. (1976)
hudsonianum	Nephrodium Brack. (1854)	Pneumatopteris Holtt. (1973)
huegelii	Polypodium Ettingsh. (1865)	= erubescens Hook.
humbertii	Pneumatopteris Holtt. (1973)	Pneumatopteris Holtt. (1973)
humilis	Sphaerostephanos Holtt. (1982)	Sphaerostephanos Holtt. (1982)
hunsteiniana	Dryopteris Brause (1920)	= fulgens
hyalostegium	Athyrium Copel. (1906)	= grammitoides
hydrophila	Phegopteris Fée (1866)	= Amauropelta
iguapensis	Dryopteris C. Chr. (1937)	Goniopteris Brade (1972)
illicita	Dryopteris Christ (1909)	= Amauropelta
imbricata	Pneumatopteris Holtt. (1973)	Pneumatopteris Holtt. (1973)
imbricatum	Polypodium Liebm. (1849)	Goniopteris Löve & Löve (1977)
imitata	Dryopteris C. Chr. (1937)	= Goniopteris
immersum	Aspidium Bl. (1828)	Amphineuron Holtt. (1974)
imponens	Polypodium Ces. (1877)	Chingia Holtt. (1974)
impressum	Nephrodium Desv. (1827)	= opulentum
inabonensis	Thelypteris Proctor (1985)	= Amauropelta
inaequans	Dryopteris C. Chr. (1913)	= Amauropelta
inaequilaterale	Nephrodium Bak. (1868)	= attenuata Brack. & hopeanum var. acutilobum
inaequilaterum	Nephrodium Colenso (1888)	= interrupta Willd.
inaequilobata	Nannothelypteris Holtt. (1971)	Nannothelypteris Holtt. (1971)
incanum	Aspidium Christ (1905)	= leprieurii var.
incerta	Dryopteris Domin (1913)	= opulentum
incisa	Pneumatopteris Holtt. (1973)	Pneumatopteris Holtt. (1973)
incisum	Polypodium Sw. (1788)	= scolopendrioides
inclusa	Dryopteris Copel. (1929)	Pneumatopteris Holtt. (1973)
inconspicua	Dryopteris Copel. (1917)	Sphaerostephanos Holtt. (1982)

Epithet	Genus in which Originally Described	Present or Proposed Disposition
indica	*Dryopteris* v.A.v.R. (1909)	= *articulatum*
indochinensis	*Dryopteris* Christ (1908)	= *hirsutipes*
indrapurae	*Sphaerostephanos* Holtt. (1982)	*Sphaerostephanos* Holtt. (1982)
induens	*Polypodium* Maxon (1905)	= *Amauropelta*
inopinata	*Coryphopteris* Holtt. (1982)	*Coryphopteris* Holtt. (1982)
insculptum	*Nephrodium* Desv. (1827)	= *unitum* (*Polypodium* L.)
insigne	*Aspidium* Mett. (1864)	*Steiropteris* Pic. Ser. (1973)
insularis	*Abacopteris* Iwats. (1959)	*Pronephrium* Holtt. (1972)
interrupta	*Pteris* Willd. (1794)	*Cyclosorus* Ito (1937)
intromissa	*Dryopteris* C. Chr. (1937)	= *Amauropelta*
invisum	*Polypodium* Forst. (1786)	*Sphaerostephanos* Holtt. (1976)
x *invisum*	*Aspidium* Sw. (1801)	*Christella* Pic. Ser. (1977)
var. *aequitorialis*	*Dryopteris* C. Chr. (1913)	= *grandis* var.
involutum	*Polypodium* Desv. (1811)	= *Amauropelta*
irayensis	*Cyclosorus* Copel. (1952)	*Sphaerostephanos* Holtt. (1982)
irenae	*Dryopteris* Brade (1965)	= *Amauropelta*
iridescens	*Dryopteris* v.A.v.R. (1913)	= *glandulosum* Bl.
isomorphus	*Sphaerostephanos* Holtt. (1982)	*Sphaerostephanos* Holtt. (1982)
iwatsukii	*Coryphopteris* Holtt. (1976)	*Coryphopteris* Holtt. (1976)
izuensis	*Dryopteris* Kodama (1913)	= *omeiense*
jacobsii	*Cyclosorus* Holtt. (1962)	= *aquatiloides*
jaculosum	*Aspidium* Christ (1904)	*Christella* Holtt. (1976)
jamaicense	*Nephrodium* Bak. (1877)	= *sanctum* var.
jamesonii	*Nephrodium* Hook. (1862)	= *Goniopteris*
janeirensis	*Dryopteris* Rosenst. (1915)	= *Amauropelta*
japenensis	*Pneumatopteris* Holtt. (1982)	*Pneumatopteris* Holtt. (1982)
japonicum	*Nephrodium* Bak. (1891)	*Parathelypteris* Ching (1963)
var. *elongata*	*Dryopteris* Rosenst. (1914)	= *chinensis*
javanica	*Mesochlaena* R. Br.ex Mett. (1856)	= *polycarpon* Bl.
jenmanii	*Nephrodium* Bak. (1877)	= *Amauropelta*
jerdonii	*Cyclosorus* Ching (1938)	= *nudatum*?
jermyi	*Pneumatopteris* Holtt. (1973)	*Pneumatopteris* Holtt. (1973)
jimenezii	*Dryopteris* Maxon & C. Chr. (1914)	= *Amauropelta*
jinfoshanensis	*Leptogramma* Ching & Liu (1984)	= *Stegnogramma*
jinfoshanensis	*Parathelypteris* Ching & Liu (1984)	*Parathelypteris* Ching & Liu (1984)
jinfoshanensis	*Stegnogramma* Ching & Liu (1983)	*Stegnogramma* Ching & Liu (1983)
jiulungshanensis	*Cyclosorus* Ching & Yao ex Ching (1982)	= *Christella*
johannae	*Pseudocyclosorus* Holtt. (1974)	*Pseudocyclosorus* Holtt. (1974)
juergensii	*Nephrodium* Rosenst. (1913)	= *Amauropelta*
jujuyensis	*Thelypteris* de la Sota (1973)	= *Amauropelta*

Epithet	Genus in which Originally Described	Present or Proposed Disposition
jungersenii	*Meniscium* Fée (1852)	*Meniscium* Fée (1852)
juruensis	*Dryopteris* C. Chr. (1912)	*Goniopteris* Brade (1972)
kalkmanii	*Sphaerostephanos* Holtt. (1982)	*Sphaerostephanos* Holtt. (1982)
kamtshatica	*Dryopteris* Komarov (1914)	= *quelpaertensis*
kaulfussii	*Aspidium* Link (1833)	= *Amauropelta*
kennedyi	*Meniscium* F.v.Muell. (1864)	= *aspera*
keraudrenianum	*Polypodium* Gaud. (1827)	*Pseudophegopteris* Holtt. (1969)
var. *tripinnata*	*Phegopteris* Hillebr. (1888)	= *procera* Mann
kerintjiensis	*Pneumatopteris* Holtt. (1973)	*Pneumatopteris* Holtt. (1973)
keysseriana	*Dryopteris* Rosenst. (1912)	*Pneumatopteris* Holtt. (1973)
khamptora	*Thelypteris* Holtt. (1971)	= *Trigonospora*
khasiensis	*Thelypteris* Ching (1936)	*Cyclogramma* Tagawa (1938)
kiauensis	*Dryopteris* C. Chr. (1937)	*Amphineuron* Holtt. (1982)
kinabaluensis	*Dryopteris* Copel. (1917)	= *gymnopodum*
kinabaluensis	*Pseudophegopteris* Holtt. (1969)	*Pseudophegopteris* Holtt. (1969)
kingii	*Thelypteris* Reed (1969)	= *leptogrammoides* Iwats.
kjellbergii	*Pronephrium* Holtt. (1972)	*Pronephrium* Holtt. (1972)
klossii	*Lastrea* Ridl. (1916)	*Coryphopteris* Holtt. (1976)
knysnaensis	*Thelypteris* N. C. Anthony & Schelpe (1982)	= *Amauropelta*
kohautiana	*Lastrea* Presl (1836)	= *unita* Kze
kolombangarae	*Coryphopteris* Holtt. (1976)	*Coryphopteris* Holtt. (1976)
koordersii	*Aspidium* Christ (1897)	*Trigonospora* Holtt. (1974)
korthalsii	*Dryopteris* Rosenst. (1917)	= *menisciicarpon*
kostermansii	*Plesioneuron* Holtt. (1982)	*Plesioneuron* Holtt. (1982)
kotoensis	*Dryopteris* Hayata (1915)	= *productum* Kaulf
krayanensis	*Thelypteris* Iwats. & Kato (1983)	= *Coryphopteris*
kuhlmannii	*Dryopteris* Brade (1964)	*Goniopteris* Brade (1972)
kumaunica	*Christella* Holtt. (1976)	*Christella* Holtt. (1976)
kundipense	*Plesioneuron* Holtt. (1975)	*Plesioneuron* Holtt. (1975)
kunthei	*Nephrodium* Desv. (1827)	= *normalis* C. Chr.
kunzeanum	*Nephrodium* Hook. (1862)	= *Christella*
kunzei	*Aspidium* Fée (1865)	= *Amauropelta*
kurzii	*Sphaerostephanos* Holtt. (1979)	*Sphaerostephanos* Holtt. (1979)
kusaiana	*Dryopteris* Hosok. (1936)	= *longissima*
kwashotensis	*Dryopteris* Hayata (1915)	= *truncatum* Poir. var.
labuanensis	*Dryopteris* C. Chr. (1905)	= *borneense*
laetestrigosum	*Nephrodium* Clarke (1880)	= *articulatum*
laevifrons	*Dryopteris* Hayata (1914)	= *truncatum* Poir. var.
laevigata	*Phegopteris* Kuhn (1869)	= *Amauropelta*
laeve	*Aspidium* Mett. (1858)	*Pneumatopteris* Holtt. (1973)
lakhimpurensis	*Dryopteris* Rosenst. (1917)	*Pronephrium* Holtt. (1972)
lamii	*Sphaerostephanos* Holtt. (1982)	*Sphaerostephanos* Holtt. (1982)
lanata	*Phegopteris* Fée (1852)	= ? *Pseudophegopteris*

Epithet	Genus in which Originally Described	Present or Proposed Disposition
lanceola	Dryopteris Christ (1907)	= rhombea
lanipes	Dryopteris C. Chr. (1909)	= Amauropelta
lanosa	Dryopteris C. Chr. (1913)	Christella Löve & Löve (1977)
larutense	Nephrodium Bedd. (1892)	Sphaerostephanos C. Chr. (1934)
lasiethes	Aspidium Kze. (1850)	= Amauropelta
lasiocarpa	Dryopteris Hayata (1911)	= torresianum
lasiopteris	Nephrodium Sod. (1883)	= Amauropelta
lastreoides	Pronephrium Presl (1851)	Sphaerostephanos Holtt. (1975)
lata	Dryopteris Hieron. (1907)	= Steiropteris
latebrosum	Aspidium Mett. (1858)	Sphaerostephanos Holtt. (1974)
laterepens	Polypodium Trotter & Hope (1899)	= pyrrhorhachis
laticuneata	Pneumatopteris Holtt. (1973)	= sogerensis
latifolium	Nephrodium Presl (1851)	= menisciicarpon
latiloba	Thelypteris Ching (1936)	Pseudocyclosorus Ching (1963)
latipinna	Nephrodium Hook. (1867)	= subpubescens Bl.
latisquamata	Pneumatopteris Holtt. (1973)	Pneumatopteris Holtt. (1973)
latum	Aspidium Mett. (1858)	= crassifolium Bl.
laui	Cyclosorus Ching (1964)	= ? siamensis
lauterbachii	Dryopteris Brause (1912)	Coryphopteris Holtt. (1976)
lawakii	Pneumatopteris Holtt. (1982)	Pneumatopteris Holtt. (1982)
laxum	Aspidium Franch. & Sav. (1879)	Metathelypteris Ching (1963)
lebeufii	Polypodium Bak. (1891)	Christella Holtt. (1974)
lechleri	Nephrodium Hieron. (1904)	= membranacea
ledermannii	Athyrium Hieron. (1920)	Coryphopteris Holtt. (1976)
leiboldiana	Lastrea Presl (1849)	= Amauropelta
lenormandii	Dryopteris C. Chr. (1906)	Cyclosorus Ching (1941)
lepida	Lastrea Moore (1886)	= Christella
lepidopodus	Cyclosorus Tard. & C. Chr. (1938)	= truncatum Poir.
lepidula	Dryopteris Hieron. (1907)	= Thelypteris s.l.
leprieurii	Nephrodium Hook. (1862)	Steiropteris Pic. Ser. (1973)
leptocladia	Goniopteris Fée (1866)	Goniopteris Fée (1866)
leptogrammoides	Dryopteris Rosenst. (1910)	= Amauropelta
leptogrammoides	Stegnogramma Iwats. (1963)	Stegnogramma Iwats. (1963)
leptoptera	Phegopteris Fée (1852)	= Amauropelta
letouzeyi	Abacopteris Tard. (1960)	= blastophorus
leucadenius	Cyclosorus Copel. (1952)	= productum Kaulf.
leucochaete	Dryopteris Slosson (1913)	= lurida
leucolepis	Lastrea Presl (1849)	= polypodioides Hook.
leuconevron	Nephrodium Fée (1852)	= unitum L.
leucophlebia	Aspidium Christ (1904)	Goniopteris Ching (1940)
leucothrix	Dryopteris C. Chr. (1909)	= Amauropelta
leveillei	Dryopteris Christ (1909)	= Cyclogramma
levingei	gymnogramme Clarke (1880)	pseudophegopteris Ching (1963) & = bukoensis

Epithet	Genus in which Originally Described	Present or Proposed Disposition
levyi	Aspidium Fourn. (1872)	= Goniopteris
l'herminieri	Aspidium Kze. (1858)	= Steiropteris
ligulata	Lastrea Presl (1851)	Pneumatopteris Holtt. (1973)
limaensis	Dryopteris Copel. (1941)	= Amauropelta
limbatum	Aspidium Sw. (1801)	Amauropelta Pic. Ser. (1977)
limbospermum	Polypodium Allioni (1789)	Oreopteris Holub (1969)
limonensis	Dryopteris Christ (1910)	= hispidulum Dcne
lindheimeri	Dryopteris C. Chr. (1913)	= augescens var.
lindigii	Dryopteris C. Chr. (1906)	= Amauropelta
lindmanii	Dryopteris C. Chr. (1907)	= Amauropelta
linearis	Dryopteris Copel. (1917)	= badia
lineatum	Aspidium Bl. (1828)	Pronephrium Presl (1851)
var. subacrostichoides	Dryopteris v.A.v.R. (1917)	= peltata
lineatum	Polypodium Coleb. (1863)	= penangianum
lingulata	Dryopteris C. Chr. (1913)	Meniscium Pic. Ser. (1968)
linkiana	Grammitis Presl (1836)	= Amauropelta
linnaeana	Lastrea Copel. (1947)	= Amauropelta
lithophila	Pneumatopteris Holtt. (1982)	Pneumatopteris Holtt. (1982)
lithophylla	Dryopteris Copel. (1917)	Sphaerostephanos Holtt. (1982)
litigiosum	Polypodium Liebm. (1849)	= Amauropelta
liukiuensis	Meniscium Christ (1910)	= ramosii
lobangensis	Dryopteris C. Chr. & Holtt. (1934)	Sphaerostephanos Holtt. (1982)
lobatus	Cyclosorus Copel. (1952)	Sphaerostephanos Holtt. (1975)
lobbiana	Goniopteris Fée (1852)	= callosum
lobulatum	Aspidium Christ (1904)	= taiwanensis
logavensis	Dryopteris Rosenst. (1912)	= ceramica
loherianum	Aspidium Christ (1893)	Sphaerostephanos Holtt. (1975)
lomatosora	Dryopteris Copel. (1941)	= Amauropelta
lonchodes	Aspidium Eaton (1860)	Steiropteris Pic. Ser. (1973)
longbawanensis	Thelypteris Iwats. & Kato (1983)	= Sphaerostephanos
longicaule	Nephrodium Bak. (1881)	= Amauropelta
longicuspe	Nephrodium Bak. (1877)	= pulchrum
longifolia	Dryopteris Bonap. (1917)	= remotipinna
longifolium	Meniscium Desv. (1827)	Meniscium Desv. (1827)
longifolium	Polypodium Roxb. (1844)	= heterocarpon
longipes	Aspidium Bl. (1828)	Pneumatopteris Holtt. (1973)
longipetiolata	Abacopteris Iwats. (1959)	Pronephrium Holtt. (1972)
longipilosum	Nephrodium Sod. (1908)	= Amauropelta
longissima	Goniopteris Brack. (1854)	Chingia Holtt. (1974)
var. novoguineensis	Dryopteris Rosenst. (1915)	= distincta
lorentzii	Aspidium Hieron. (1897)	= argentinum
loretensis	Dryopteris Maxon (ined.)	= Amauropelta
lorzingii	Chingia Holtt. (1982)	Chingia Holtt. (1982)
lowei	Gymnogramme Hook. & Grev. (1829)	= pozoi

Epithet	Genus in which Originally Described	Present or Proposed Disposition
lucbanii	*Sphaerostephanos* Holtt. (1982)	*Sphaerostephanos* Holtt. (1982)
lucidum	*Nephrodium* Bak. (1877)	*Pneumatopteris* Holtt. (1973)
luerssenii	*Nephrodium* Harr. (1877)	= *ligulata*
lugubre	*Aspidium* Mett. (1853)	*Goniopteris* Brade (1972)
lugubriformis	*Dryopteris* Rosenst. (1909)	= *Goniopteris*
luoquingensis	*Cyclosorus* Ching & Zhang (1983)	= *Thelypteris* s.l.
lurida	*Dryopteris* Slosson (1913)	= *Thelypteris* s.l.
lustratum	*Nephrodium* Hieron. (1904)	= *Amauropelta*
luxurians	*Polypodium* Kze. (1850)	= *prolifera*
luzonica	*Dryopteris* Christ (1907)	= *laeve*
var. *puberula*	*Dryopteris* Christ (1907)	= *productum* Kaulf.
macarthyi	*Nephrodium* Bak. (1891)	= *laxum*
macbridei	*Dryopteris* Maxon (1944)	= *Amauropelta*
macgregori	*Nephrodium* Bak. (1892)	= *fasciculatum*
macilenta	*Thelypteris* E.P. St.John (1936)	= *Christella*
macra	*Thelypteris* A. R. Smith (1983)	= *Amauropelta*
macradenium	*Nephrodium* Sod. (1883)	= *Amauropelta*
macrophyllum	*Meniscium* Kze. (1839)	*Meniscium* Kze (1839)
macrophyllus	*Cyclosorus* Ching & Liu (1983)	= *Christella*?
macropodum	*Aspidium* Mett. (1859)	*Goniopteris* Brade (1972)
macroptera	*Dryopteris* Copel. (1931)	*Pneumatopteris* Holtt. (1973)
macropus	*Aspidium* Mett. (1859)	*Goniopteris* Ching (1940)
macrorhizoma	*Thelypteris* E.P. St.John (1943)	= *Christella*
macrotis	*Gymnogramme* Kze. (1848)	= *affine* Bl.
macrotis	*Nephrodium* Hook. (1862)	*Goniopteris* Pic. Ser. (1977)
macrourum	*Aspidium* Kaulf. (1823)	= *patens* Sw.
madagascariensis	*Goniopteris* Fée (1852)	= *unita* Kze
maemonensis	*Cyclosorus* Wagner & Greth. (1948)	*Sphaerostephanos* Holtt. (1977)
magdalenica	*Dryopteris* Hieron. (1907)	= *Goniopteris*
magnifica	*Dryopteris* Copel. (1929)	*Pneumatopteris* Holtt. (1973)
magnus	*Cyclosorus* Copel. (1952)	*Sphaerostephanos* Holtt. (1975)
majus	*Haplodictyum* Copel. (1920)	*Sphaerostephanos* Holtt. (1982)
makassaricus	*Sphaerostephanos* Holtt. (1982)	*Sphaerostephanos* Holtt. (1982)
malabariense	*Nephrodium* Fée (1865)	*Christella* Holtt. (1976)
malaccensis	*Lastrea* Presl (1851)	= *opulentum*
malacothrix	*Dryopteris* Maxon (1930)	= *asterothrix*
malangae	*Dryopteris* C. Chr. (1931)	= *Amauropelta*
malayensis	*Dryopteris* C. Chr. (1913)	= *glandulosum* Bl.
malodora	*Dryopteris* Copel. (1936)	*Chingia* Holtt. (1974)
mamberamensis	*Phegopteris* v.A.v.R. (1917)	= *ceramica*
manilensis	*Lastrea* Presl (1851)	= *Ctenitis*

Epithet	Genus in which Originally Described	Present or Proposed Disposition
mapiriensis	*Dryopteris* Rosenst. (1909)	= *decussatum* var.
maquilingensis	*Dryopteris* Copel. (1935)	= *granulosum*
maranguense	*Aspidium* Hieron. (1895)	= *bergianum*
marattioides	*Dryopteris* Alston (1940)	*Plesioneuron* Holtt. (1975)
margaretae	*Dryopteris* E. Brown (1931)	*Amauropelta* Holtt. (1977)
marginalis	*Polypodium* L. (1753)	= *Dryopteris*
marginifera	*Hypopeltis* Bory (1833)	= *interrupta* Willd.
marquesicum	*Plesioneuron* Holtt. (1977)	*Plesioneuron* Holtt. (1977)
marthae	*Dryopteris* v.A.v.R. (1911)	= *calcaratum* Bl.
martinii	*Dryopteris* C. Chr. (1906)	= *Thelypteris* s.l.
mascarense	*Polypodium* Bak. (1868)	= *distans* Hook.
mataanae	*Dryopteris* Brause (1922)	= *pubirachis*
matutumensis	*Dryopteris* Copel. (1929)	= *urdanetensis*
mauiensis	*Dryopteris* C. Chr. (1906)	= *Ctenitis*
mauritiana	*Dryopteris* C. Chr. (1906)	= *elatum Boj.*
mauritianum	*Nephrodium* Fée (1852)	= *elatum Boj.*
var. *gardineri*	*Dryopteris* C. Chr. (1912)	= *subtruncatum*
maxima	*Thelypteris* Iwats. & Kato (1983)	= *Sphaerostephanos*
maximum	*Aspidium* Bak. (1874)	= *elatum Boj.*
media	*Dryopteris* v.A.v.R. (1913)	= *singalanense* Bak.
medogensis	*Cyclosorus* Ching & Wu (1983)	= *Christella*
medogensis	*Pseudocyclosorus* Ching & Wu (1983)	*Pseudocyclosorus* Ching & Wu (1983)
medusella	*Plesioneuron* Holtt. (1982)	*Plesioneuron* Holtt. (1982)
meeboldii	*Dryopteris* Rosenst. (1913)	= *malabariense*
megacuspe	*Polypodium* Bak. (1890)	*Pronephrium* Holtt. (1973)
megalocarpa	*Dryopteris* v.A.v.R. (1922)	= *hirsutipes*
megalodus	*Polypodium* Schkuhr (1806)	= *Goniopteris*
megaphylloides	*Dryopteris* v.A.v.R. (1915)	= *pseudomegaphylla*
megaphylloides	*Dryopteris* Rosenst. (1913)	= *pilososquamata*
megaphyllum	*Aspidium* Mett. (1864)	= *pennigerum* Hook.
var. *hirsutum*	*Aspidium* Mett. (1864)	= *pseudomegaphylla*
meiobasis	*Coryphopteris* Holtt. (1976)	*Coryphopteris* Holtt. (1976)
melanochlaena	*Dryopteris* C. Chr. (1909)	= *Amauropelta*
melanophlebia	*Dryopteris* Copel. (1911)	*Pronephrium* Holtt. (1972)
melanorachis	*Sphaerostephanos* Holtt. (1982)	*Sphaerostephanos* Holtt. (1982)
melanorhiza	*Aspidium* Christ (1901)	= *flexile*
membranacea	*Phegopteris* Mett. (1859)	*Meniscium* Pic. Ser. (1968)
membranifera	*Dryopteris* C. Chr. (1925)	*Amauropelta* Holtt. (1974)
menadensis	*Sphaerostephanos* Holtt. (1982)	*Sphaerostephanos* Holtt. (1982)
mengenianus	*Sphaerostephanos* Holtt. (1982)	*Sphaerostephanos* Holtt. (1982)
menisciicarpon	*Aspidium* Bl. (1828)	*Pronephrium* Holtt. (1972)
meniscioides	*Goniopteris* Fée (1852)	= *prolifera*
meniscioides	*Polypodium* Liebm. (1849)	*Goniopteris* Fourn. (1872)
mercurii	*Dryopteris* Hieron. (1907)	= *Amauropelta*
merillii	*Dryopteris* Christ (1907)	*Pronephrium* Holtt. (1972)
mertensioides	*Dryopteris* C. Chr. (1907)	= *Amauropelta*

25

Epithet	Genus in which Originally Described	Present or Proposed Disposition
mesocarpa	*Dryopteris* Copel. (1932)	*Pneumatopteris* Holtt. (1973)
metcalfei	*Polypodium* Bak. (1891)	*Sphaerostephanos* Holtt. (1977)
metteniana	*Thelypteris* Ching (1941)	= *Amauropelta*
mettenii	*Lastrea* Copel. (1947)	= *Christella*
mexiae	*Dryopteris* Copel. (1932)	= *Steiropteris*
micans	*Dryopteris* Brause (1920)	= *rudis* var.
michaelis	*Pneumatopteris* Holtt. (1973)	*Pneumatopteris* Holtt. (1973)
microauriculata	*Pneumatopteris* Holtt. (1973)	*Pneumatopteris* Holtt. (1973)
microbasis	*Nephrodium* Bak. (1874)	*Christella* Holtt. (1974)
microcarpa	*Dryopteris* v.A.v.R. (1920)	= *beddomei*
microcarpon	*Gymnogramme* Fée (1857)	= *decussatum*
microcarpon	*Nephrodium* Fée (1852)	= *unitum* L.
microcarpum	*Nephrodium* Presl (1836)	= *interrupta* Willd.
microchlamys	*Lastrea* de Vriese (1848)	= *polycarpon* Bl.
microchlamys	*Nephrodium* Bak. (1876)	= *polycarpon* Bl.
microdendron	*Polypodium* Eaton (1868)	= *sandwicensis*
microlepigera	*Coryphopteris* Holtt. (1976)	*Coryphopteris* Holtt. (1976)
microloncha	*Dryopteris* Christ (1907)	*Pneumatopteris* Holtt. (1973)
micropaleata	*Pneumatopteris* Holtt. (1973)	*Pneumatopteris* Holtt. (1973)
micropinnatum	*Pronephrium* Holtt. (1972)	*Pronephrium* Holtt. (1972)
microsora	*Dryopteris* Copel. (1929)	= *parksii*
microsorum	*Nephordium* Clarke (1880)	= *appendiculatum* Presl
microstegium	*Nephrodium* Hook. (1862)	*Pseudophegopteris* Ching (1983)
millarae	*Pronephrium* Holtt. (1972)	*Pronephrium* Holtt. (1972)
millei	*Dryopteris* C. Chr. (1913)	= *Amauropelta*
minahassae	*Pronephrium* Holtt. (1972)	*Pronephrium* Holtt. (1972)
mindanaensis	*Dryopteris* Christ (1907)	= *dentatum*
mindorensis	*Sphaerostephanos* Holtt. (1975)	*Sphaerostephanos* Holtt. (1975)
minensis	*Thelypteris* Abbiat. (1964)	= *Amauropelta*
mingchegensis	*Dictyocline* Ching (1963)	= *Stegnogramma*
mingendensis	*Lastrea* Gilli (1978)	*Pneumatopteris* Holtt. (1982)
minima	*Christella* Holtt. (1976)	*Christella* Holtt. (1976)
minor	*Dryopteris* C. Chr. (1913)	= *Goniopteris*
minuscula	*Dryopteris* Maxon (1932)	*Meniscium* Pic. Ser. (1968)
minutula	*Thelypteris* Morton (1953)	= *Amauropelta*
miqueliana	*Lastrea* Tagawa (1953)	= *angustifrons*
mirabilis	*Dryopteris* Copel. (1911)	= *menisciicarpon*
mixta	*Dryopteris* Rosenst. (1913)	= *heterocarpon*
miyagii	*Dryopteris* Ito (1935)	= *glanduligerum*
mjobergii	*Sphaerostephanos* Holtt. (1982)	*Sphaerostephanos* Holtt. (1982)
modesta	*Christella* Holtt. (1974)	*Christella* Holtt. (1974)
molle	*Aspidium* Sw. (1801)	= *dentatum*
var. *aureum*	*Nephrodium* Clarke (1883)	= *cylindrothrix*
var. *decurtatum*	*Aspidium* Bailey (1892)	= *dentatum* var.
var. *keffordii*	*Aspidium* Bailey (1892)	= *dentatum* var.
var. *major*	*Nephrodium* Bedd. (1892)	= *papilio*
molle	*Nephrodium* sensu Hook., Sp. Fil. (1862)	= *heterocarpum*

Epithet	Genus in which Originally Described	Present or Proposed Disposition
molle	Polypodium Jacq. (1989)	= dentatum
mollicella	Dryopteris Maxon (1923)	= Amauropelta
molliculum	Polypodium Kze (1841)	= Amauropelta
mollis	Phegopteris Kuhn (1868)	= dianae
mollis	Phegopteris Mett. (1854)	= Meniscium
mollissima	Gymnogramme Fisch. (1850)	= pozoi
mollissimum	Aspidium Christ (1898)	= flaccidum Bl.
molliusculum	Aspidium Kuhn (1868)	= canum
molliusculum	Nephrodium Bedd. (1892)	= appendiculata Bl.
moluccana	Dryopteris C. Chr. (1937)	= ceramica
molundensis	Dryopteris Brause (1915)	= striatum var.
mombachensis	Thelypteris L. Gómez (1982)	= Amauropelta
moniliformis	Dimorphopteris Tagawa & Iwats. (1961)	Pronephrium Holtt. (1972)
monodonta	Dryopteris C. Chr. (1906)	= unidentata
monosora	Lastrea Presl (1849)	Goniopteris Brade (1972)
montbrisoniana	Phegopteris Fée (1952)	= aubertii
moritziana	Dryopteris Urban (1903)	= Amauropelta
morobensis	Dryopteris Copel. (1942)	= dimorpha Brause
morotaiensis	Sphaerostephanos Holtt. (1982)	Sphaerostephanos Holtt. (1982)
mortonii	Thelypteris A. R. Smith (1973)	= Amauropelta
moselyi	Sphaerostephanos Holtt. (1982)	Sphaerostephanos Holtt. (1982)
mosenii	Dryopteris C. Chr. (1907)	= Amauropelta
motleyanum	Nephrodium Hook. (1867)	Mesophlebion Holtt. (1974)
moulmeinense	Nephrodium Bedd. (1876)	= nudatum
moussetii	Dryopteris Rosenst. (1910)	= rectangulare
mucosa	Thelypteris A. R. Smith (1973)	Amauropelta Löve & Löve (1977)
multiauriculata	Dryopteris Copel. (1942)	Sphaerostephanos Holtt. (1982)
multiformis	Dryopteris C. Chr. (1913)	= Amauropelta
multifrons	Dryopteris C. Chr. (1925)	= Christella
multijugum	Nephrodium Bak. (1867)	= Thelypteris s.l.
multijugum	Aspidium Christ (1898)	= callosum
multilineata	Goniopteris Bedd. (1867)	= pennigerum Hook.
multilineatum	Aspidium Mett. (1858)	= unitum (Polypodium L.) var.
multilineatum	Polypodium Hook. (1844)	= nudatum
var. malayensis	Abacopteris Holtt. (1955)	= aspera
var. assamicum	Nephrodium Bedd. (1893)	= subelatum
multisetum	Nephrodium Bak. (1886)	Macrothelypteris Ching (1963)
multisora	Dryopteris C. Chr. & Holtt. (1934)	Coryphopteris Holtt. (1976)
muluensis	Sphaerostephanos Holtt. (1982)	Sphaerostephanos Holtt. (1982)
munchii	Thelypteris A. R. Smith (1973)	= Goniopteris

Epithet	Genus in which Originally Described	Present or Proposed Disposition
munda	*Dryopteris* Rosenst. (1917)	*Sphaerostephanos* Holtt. (1982)
muricata	*Dryopteris* Brause (1920)	*Chingia* Holtt. (1974)
var. *marginata*	*Dryopteris* Brause (1920)	= *imponens*
var. *obscura*	*Dryopteris* Brause (1920)	= *imponens*
muricatum	*Polypodium* Powell (1874)	= *glandulifera* Brack.
musicola	*Thelypteris* Proctor (1961)	= *Amauropelta*
mutabilis	*Dryopteris* Brause (1920)	*Sphaerostephanos* Holtt. (1982)
muzensis	*Dryopteris* Hieron. (1907)	= *Amauropelta*
myriosora	*Dryopteris* Copel. (1936)	*Plesioneuron* Holtt. (1975)
nakaikei	*Sphaerostephanos* Holtt. (1982)	*Sphaerostephanos* Holtt. (1982)
namaphila	*Thelypteris* Proctor (1985)	= *Amauropelta*
namburenense	*Nephrodium* Bedd. (1892)	*Christella* Holtt. (1974)
nana	*Christella* Holtt. (1976)	*Christella* Holtt. (1976)
nanchuanensis	*Cyclosorus* Ching & Liu (1983)	= *Christella*
natalense	*Aspidium* Fée (1857)	?= *dentatum*
navarrense	*Aspidium* Christ (1907)	*Amauropelta* Pic. Ser. (1977)
neglecta	*Dryopteris* Brade & Rosenst. (1931)	= *Amauropelta*
negligens	*Nephrodium* Jenm. (1896)	= *Amauropelta*
nemorale	*Nephrodium* Sod. (1893)	= *Goniopteris*
nemorale	*Polypodium* Brack. (1854)	= *torresianum*
nemoralis	*Thelypteris* Ching (1936)	= *Metathelypteris*
neoauriculata	*Dryopteris* Ching (1931)	*Cyclogramma* Tagawa (1938)
neotoppingii	*Sphaerostephanos* Holtt. (1982)	*Sphaerostephanos* Holtt. (1982)
nephrodioides	*Aspidium* Klotzsch (1847)	*Goniopteris* Brade (1972)
nephrodioides	*Aspidium* Hook. (1862)	= *opulentum*
nephrodioides	*Aspidium* Fée (1869)	= *patens* Sw. var. *dissimilis*
nephrodioides	*Lastrea* Bedd. (1866)	= *chlamydophora*
nephrolepioides	*Dryopteris* C. Chr. (1937)	*Pneumatopteris* Holtt. (1973)
nervosa	*Phegopteris* Fée (1852)	*Nannothelypteris* Holtt. (1973)
nervosum	*Polypodium* Klotzsch (1847)	= *Amauropelta*
nesiotica	*Dryopteris* Maxon & Morton (1938)	*Meniscium* Pic. Ser. (1968)
nevadense	*Nephrodium* Bak. (1891)	*Parathelypteris* Holtt. (1976)
nicaraguensis	*Phegopteris* Fourn. (1872)	= *Goniopteris*
nigrescentium	*Polypodium* Jenm. (1895)	*Goniopteris* Ching (1940)
nigricans	*Dryopteris* Ekm. & C. Chr. (1937)	= *Goniopteris*
nimbatum	*Nephrodium* Jenm. (1894)	= *Amauropelta*
nipponicum	*Aspidium* Franch. & Sav. (1867)	*Parathelypteris* Ching (1963)
nitens	*Polypodium* Desv. (1827)	= *Amauropelta*
nitidum	*Pronephrium* Holtt. (1972)	*Pronephrium* Holtt. (1972)
nitidulum	*Nephrodium* Presl (1851)	*Pneumatopteris* Holtt. (1973)
nockianum	*Nephrodium* Jenm. (1886)	*Amauropelta* Pic. Ser. (1977)

Epithet	Genus in which Originally Described	Present or Proposed Disposition
normalis	Dryopteris C. Chr. (1910)	Christella Holtt. (1976)
normalis	Lastrea Copel. (1947)	= resiniferum Desv.
norrisii	Dryopteris Rosenst. (1917)	Sphaerostephanos Holtt. (1982)
notabilis	Dryopteris Brause (1920)	Plesioneuron Holtt. (1975)
nothofagetii	Pronephrium Holtt. (1972)	= adenostegia
novaeana	Dryopteris Brade (1936)	= Amauropelta
novae-britanniae	Sphaerostephanos Holtt. (1982)	Sphaerostephanos Holtt. (1982)
novae-caledoniae	Pneumatopteris Holtt. (1973)	Pneumatopteris Holtt. (1973)
novae-hiberniae	Thelypteris Holtt. (1967)	= harveyi
noveboracensis	Polypodium L. (1753)	Parathelypteris Ching (1963)
novoguineensis	Dryopteris Brause (1912)	Sphaerostephanos Holtt. (1982)
nubicola	Thelypteris de la Sota (1973)	= Amauropelta
nubigena	Thelypteris A. R. Smith (1975)	= Amauropelta
nudatum	Polypodium Roxb. (1844)	Pronephrium Holtt. (1972)
nudisorus	Sphaerostephanos Holtt. (1982)	Sphaerostephanos Holtt. (1982)
nuna	Dryopteris J. W. Moore (1933)	= grantii
nymphale	Polypodium Forst. (1786)	= dentatum
oaxacana	Thelypteris A. R. Smith (1973)	= Amauropelta
oblanceolata	Dryopteris Copel. (1914)	= beccarianum (Meniscium Ces.)
oblancifolia	Dryopteris Tagawa (1936)	= dentatum
obliqua	Pneumatopteris Holtt. (1973)	Pneumatopteris Holtt. (1973)
obliquatum	Aspidium Mett. (1861)	= prolixa Willd.
var. germinyi	Aspidium Linden (1880)	= harveyi var. connivens
obliteratum	Polypodium Sw. (1788)	Goniopteris Presl (1836)
oblonga	Dryopteris Brause (1920)	= munda
obscurum	Aspidium Bl. (1828)	= aridum
obstructa	Dryopteris Copel. (1931)	Pneumatopteris Holtt. (1973)
obtusata	Dryopteris v.A.v.R. (1918)	Coryphopteris Holtt. (1976)
obtusatum	Aspidium Sw. (1800)	= interrupta Willd.
obtusifolia	Dryopteris Rosenst. (1912)	Sphaerostephanos Holtt. (1982)
obtusiloba	Trigonospora Sledge (1981)	Trigonospora Sledge (1981)
occultum	Nephrodium Hope (1899)	= ? penangianum
ochropteroides	Nephrodium Bak. (1891)	= Thelypteris s.l.
ochthodes	Aspidium Kze. (1851)	Pseudocyclosorus Holtt. (1974)
odontosora	Dryopteris Bonap. (1917)	Amauropelta Holtt. (1976)
ogasawarensis	Dryopteris Nakai (1929)	Macrothelypteris Holtt. (1969)
ogatana	Dryopteris Koidz. (1925)	= acuminatum Houtt. var.
okinawensis	Dryopteris Ito (1935)	= angustifrons
oligocarpum	Polypodium Willd. (1810)	Amauropelta Pic. Ser. (1977)
oligodictyon	Acrostichum Bak. (1887)	Mesophlebion Holtt. (1975)

29

Epithet	Genus in which Originally Described	Present or Proposed Disposition
oligolepia	*Dryopteris* v.A.v.R. (1924)	*Coryphopteris* Holtt. (1976)
oligophlebium	*Nephrodium* Bak. (1875)	= *torresianum* var. *calvatum*
var. *lasiocarpa*	*Dryopteris* Hayata (1911)	= *torresianum*
var. *subtripinnata*	*Dryopteris* Tagawa (1933)	= *viridifrons*
oligophylla	*Dryopteris* Maxon (1908)	= *grandis*
oligophyllum	*Meniscium* Bak. (1891)	= *guyanensis*
oligosorum	*Polypodium* Klotzsch (1847)	= *Amauropelta*
omatianus	*Sphaerostephanos* Holtt. (1982)	*Sphaerostephanos* Holtt. (1982)
omeiense	*Polypodium* Bak. (1888)	*Cyclogramma* Tagawa (1938)
omeigensis	*Cyclosorus* Ching (1949)	*Christella* Holtt. (1976)
oochlamys	*Dryopteris* C. Chr. (1906)	= *kunzei*
oosorum	*Nephrodium* Bak. (1896)	*Sphaerostephanos* Holtt. (1982)
ophiura	*Dryopteris* Copel. (1942)	*Plesioneuron* Holtt. (1975)
oppositans	*Gymnogramme* Fée (1869)	= *Amauropelta*
oppositifolium	*Polypodium* Hook. (1863)	*Pneumatopteris* Holtt. (1973)
oppositiformis	*Dryopteris* C. Chr. (1925)	*Amauropelta* Holtt. (1974)
oppositipinna	*Phegopteris* v.A.v.R. (1914)	= *rectangulare*
oppositipinnus	*Cyclosorus* Ching & Liu (1984)	= *Pneumatopteris ?*
oppositum	*Polypodium* Vahl (1807)	*Amauropelta* Pic. Ser. (1977)
opulentum	*Aspidium* Kaulf. (1824)	*Amphineuron* Holtt. (1971)
var. *hirsutus*	*Cyclosorus* Ching (1938)	= *terminans*
orbicularis	*Dryopteris* C. Chr. (1906)	= *opulentum*
oregana	*Dryopteris* C. Chr. (1906)	= *nevadense*
oreopteris	*Nephrodium* Fée (1852)	= *terminans*
oreopteris	*Thelypteris* Ehrh. (1787)	= *limbospermum*
organensis	*Dryopteris* Rosenst. (1924)	= *Amauropelta*
orizabae	*Aspidium* Fée (1857)	= *Christella*
ornatipes	*Pseudocyclosorus* Holtt. & Grimes (1980)	*Pseudocyclosorus* Holtt. & Grimes (1980)
ornatum	*Polypodium* Bedd. (1864)	*Macrothelypteris* Ching (1963)
oroniensis	*Thelypteris* L. Gómez (1978)	= *Goniopteris*
orthocaulis	*Cyclosorus* Iwats. (1963)	= *decadens*
oshimense	*Aspidium* Christ (1901)	= *acuminatum* Houtt.
otaria	*Aspidium* Kze (1858)	= *Anisocampium*
ovata	*Thelypteris* R.P. St. John (1938)	*Christella* Löve & Löve (1977)
oxyotis	*Dryopteris* Rosenst. (1917)	= *affine* Bl.
oxyoura	*Dryopteris* Copel. (1936)	*Pneumatopteris* Holtt. (1973)
pabstii	*Dryopteris* Brade (1965)	= *Amauropelta*
pachyrachis	*Aspidium* Mett. (1858)	= *Amauropelta*
pachysora	*Dryopteris* Hieron. (1907)	= *Meniscium*
pacifica	*Christella* Holtt. (1976)	*Christella* Holtt. (1976)
pacifica	*Thelypteris* Reed (1968)	= *Ctenitis*
palauense	*Glaphyropteris* Hosok. (1942)	*Glaphyropteris* Hosok. (1942)
palauense	*Meniscium* Hosok. (1938)	*Pronephrium* Holtt. (1972)
palawanensis	*Thelypteris* Reed (1968)	= *granulosum*

Epithet	Genus in which Originally Described	Present or Proposed Disposition
paleacea	Thelypteris A. R. Smith (1983)	= Amauropelta
paleata	Dryopteris Copel. (1914)	= trichopoda
pallescens	Dryopteris Brause (1920)	= beccarianum (Nephrodium Ces.)
pallida	Thelypteris Ching (1941)	Pseudophegopteris Ching (1963)
pallidum	Aspidium Fourn. (1872)	= lanosa
pallidum	Polypodium Brack. (1854)	= polypodioides Hook.
pallidivenium	Polypodium Hook. (1863)	= striatum
palmii	Dryopteris C. Chr. (1916)	= bergianum
palopense	Pronephrium Holtt. (1972)	= amboinense Willd.
paludosa	Nephrodium Liebm. (1849)	= Cyclosorus
paludosum	Aspidium Bl. (1928)	Pseudophegopteris Ching (1963)
paludosum	Polypodium Bedd. (1863)	= pyrrhorhachis
palustre	Nephrodium Bak. (1867)	= mettenii
palustris	Thelypteris Schott (1834)	Thelypteris Schott (1834)
panamense	Nephrodium Presl (1825)	= resiniferum Desv.
papilio	Nephrodium Hope (1899)	Christella Holtt. (1974)
papuana	Pneumatopteris Holtt. (1973)	Pneumatopteris Holtt. (1973)
papyraceum	Nephrodium Bedd. (1892)	Christella Holtt. (1974)
paraphysata	Dryopteris Copel. (1911)	= pilososquamata
paraphysophora	Dryopteris v.A.v.R. (1920)	Amphineuron Holtt. (1977)
parasiticum	Polypodium L. (1753)	Christella Lév. (1915)
var. aureum	Nephrodium Clarke (1880)	= cylindrothrix
var. coriacea	Dryopteris Bonap. (1917)	= dentatum var.
var. falcatula	Dryopteris Christ (1907)	= hispidulum Dcne
var. multijugum	Nephrodium Clarke (1880)	= opulentum
parathelypteris	Aspidium Christ (1905)	= chinensis
paripinnata	Dryopteris Copel. (1942)	Sphaerostephanos Holtt. (1982)
parishii	Meniscium Bedd. (1866)	Pronephrium Holtt. (1972)
parksii	Dryopteris Ballard (1973)	Pneumatopteris Holtt. (1973)
patens	Goniopteris Fée (1852)	= unita Kze
patens	Polypodium Sw. (1788)	Christella Holtt. (1976)
patentipinna	Pneumatopteris Holtt. (1973)	Pneumatopteris Holtt. (1973)
patula	Gymnogramme Fée (1869)	= Amauropelta
pauciflorum	Meniscium Hook. (1864)	Menisorus Alston (1956)
paucijugum	Aspidium Klotzsch (1847)	= Goniopteris
paucijugum	Nephrodium Jenm. (1886)	= Christella
paucinervata	Dryopteris C. Chr. (1906)	= Stigmatopteris alloeoptera
paucipaleata	Chingia Holtt. (1974)	Chingia Holtt. (1974)
paucipinnatum	Nephrodium Donn.-Sm. (1887)	= Goniopteris
pauciserratus	Cyclosorus Ching & Zhang (1983)	= ? Christella
pauper	Aspidium Fée (1857)	= ? patens Sw.
pavonianum	Polypodium Klotzsch (1847)	= Amauropelta
pectiniformis	Dryopteris C. Chr. (1929)	Coryphopteris Holtt. (1976)
peekelii	Dryopteris v.A.v.R. (1908)	Christella Holtt. (1976)
peliliuensis	Thelypteris Fosberg (1980)	= palauense Hosok. (1942)

Epithet	Genus in which Originally Described	Present or Proposed Disposition
peltata	Dryopteris v.A.v.R. (1914)	Pronephrium Holtt. (1982)
peltochlamys	Dryopteris C. Chr. (1937)	Sphaerostephanos Holtt. (1982)
penangianum	Polypodium Hook. (1863)	Pronephrium Holtt. (1972)
pennatum	Polypodium Poir. (1804)	Goniopteris Pic. Ser. (1977)
pennellii	Thelypteris A. R. Smith (1980)	= Steiropteris
pennigerum	Nephrodium Hook. (1862)	Sphaerostephanos Holtt. (1974)
var. malayense	Nephrodium Bedd. (1892)	= norrisii
pennigerum	Polypodium Forst. (1786)	Pneumatopteris Holtt. (1973)
pentaphylla	Dryopteris Rosenst. (1913)	Pronephrium Holtt. (1972)
perakense	Aspidium Bedd. (1888)	= polycarpon Bl.
var. sumatrensis	Dryopteris v.A.v.R. (1913)	= appendiculata Bl.
peramalense	Pronephrium Holtt. (1972)	Pronephrium Holtt. (1972)
percivalii	Polypodium Jenm. (1891)	= decussatum
pergamacea	Pneumatopteris Holtt. (1973)	Pneumatopteris Holtt. (1973)
perglandulifera	Dryopteris v.A.v.R. (1920)	Sphaerostephanos Holtt. (1975)
peripae	Nephrodium Sod. (1883)	= Goniopteris
permollis	Dryopteris Maxon & Morton (1938)	= Meniscium
perpilifera	Dryopteris v.A.v.R. (1913)	= acrostichoides Desv.
perpubescens	Dryopteris Alston (1940)	Christella Holtt. (1976)
perrigida	Phegopteris v.A.v.R. (1914)	Chingia Holtt. (1974)
persimile	Polypodium Bak. (1876)	Pseudophegopteris Holtt. (1969)
persquamifera	Dryopteris v.A.v.R. (1914)	Mesophlebion Holtt. (1975)
perstrigosa	Dryopteris Maxon (1932)	= Amauropelta
peruviana	Dryopteris Rosenst. (1909)	= Amauropelta
petelotii	Thelypteris Ching (1936)	Coryphopteris Holtt. (1976)
petiolata	Leptogramma Ching (1963)	= pozoi var.
petiolata	Thelypteris Iwats. & Kato (1983)	= Sphaerostephanos
petrophila	Dryopteris Copel. (1942)	Pneumatopteris Holtt. (1973)
phacelothrix	Dryopteris Rosenst. (1912)	= Amauropelta
phanerophlebium	Nephrodium Bak. (1874)	Plesioneuron Holtt. (1975)
phegopteris	Polypodium L. (1753)	= connectile
philippina	Lastrea Presl (1851)	= ligulata
philippinarum	Abacopteris Fée (1852)	= aspera
philippinense	Nephrodium Bak. (1891)	= productum Kaulf.
philippinum	Physematium Presl (1851)	Nannothelypteris Holtt. (1973)
physematioides	Aspidium Krug (1897)	= Amauropelta
piedrensis	Dryopteris C. Chr. (1909)	= Amauropelta
pilosa	Gymnogramme Martens & Gall. (1842)	Stegnogramma Iwats. (1964)
pilosissima	Thelypteris Morton (1951)	= Thelypteris s.l.
pilosissimus	Sphaerostephanos Holtt. (1982)	Sphaerostephanos Holtt. (1982)
pilosiusculum	Acrostichum Wikstr. (1825)	= pozoi
pilosiusculum	Aspidium Mett. (1864	= appendiculata Bl.

Epithet	Genus in which Originally Described	Present or Proposed Disposition
pilosiusculum	Nephrodium Racib. (1898)	= debile Bak.
pilosohispidum	Nephrodium Hook. (1862)	= Amauropelta
pilososquamata	Dryopteris v.A.v.R. (1908)	Sphaerostephanos Holtt. (1982)
var. obtusata	Dryopteris v.A.v.R. (1914)	= eminens
pilosulum	Aspidium Mett. (1850)	Amauropelta Löve & Löve (1977)
pinnata	Dryopteris Copel. (1929)	= clliatum Benth.
pinnatifida	Thelypteris A. R. Smith (1983)	= Goniopteris
pinwillii	Polypodium Bak. (1891)	= repanda
pittieri	Dryopteris C. Chr. (1909)	= Amauropelta
plantianum	Nephrodium Pappe & Rawson (1858)	= interrupta Willd.
platensis	Thelypteris Abbiat. (1964)	= Amauropelta
platylobum	Plesioneuron Holtt. (1975)	Plesioneuron Holtt. (1975)
platyptera	Dryopteris Copel. (1942)	Coryphopteris Holtt. (1976)
platyrachis	Aspidium Fée (1873)	= Amauropelta
plumieri	Polypodium Desv. (1811)	= Amauropelta
plumosa	Dryopteris C. Chr. (1937)	Coryphopteris Holtt. (1976)
plurifolia	Dryopteris v.A.v.R. (1922)	Sphaerostephanos Holtt. (1982)
plurivenosus	Sphaerostephanos Holtt. (1982)	Sphaerostephanos Holtt. (1982)
pohlianum	Aspidium Presl (1822)	= interrupta Willd.
poiteana	Lastrea Bory (1826)	Goniopteris Ching (1940)
polisianus	Sphaerostephanos Holtt. (1975)	Sphaerostephanos Holtt. (1975)
polycarpon	Aspidium Bl. (1828)	Sphaerostephanos Copel (1929)
polycarpon	Polypodium Hook. & Arn. (1832)	= sandwicensis
var. kauaiensis	Phegopteris Hillebr. (1888)	= cyatheoides
polyotis	Dryopteris C. Chr. (1933)	Sphaerostephanos Holtt. (1982)
polyphlebia	Dryopteris C. Chr. (1913)	= Steiropteris
polypodioides	Alsophila Hook. (1835)	Macrothelypteris Holtt. (1969)
polypodioides	Ceterach Raddi (1819)	= Stegnogramma
polypodioides	Phegopteris Fée (1852)	= connectile
polypodioides	Pteris Poir. (1804)	= interrupta Willd.
polyphylla	Dryopteris Copel. (1941)	= Amauropelta
polypterus	Cyclosorus Copel. (1955)	= latebrosum
polytrichum	Nephrodium Bak. (1891)	= trichopoda
polytrichum	Nephrodium Schrad. (1824)	= conspersum
ponapeana	Phegopteris Hosok. (1936)	Plesioneuron Holtt. (1975)
porphyricola	Dryopteris Copel. (1912)	Sphaerostephanos Holtt. (1975)
porphyrophlebium	Aspidium Christ (1904)	= penangianum
posthumii	Sphaerostephanos Holtt. (1982)	Sphaerostephanos Holtt. (1982)

Epithet	Genus in which Originally Described	Present or Proposed Disposition
potamios	*Sphaerostephanos* Holtt. (1982)	*Sphaerostephanos* Holtt. (1982)
pozoi	*Hemionitis* Lagasca (1816)	*Stegnogramma* Iwats. (1963)
praetermissa	*Dryopteris* Maxon (1944)	= *Goniopteris*
praetervisum	*Aspidium* Kuhn (1864)	*Steiropteris* Pic. Ser. (1973)
prenticei	*Lastrea* Carr. (1873)	*Plesioneuron* Holtt. (1975)
presliana	*Dryopteris* Ching (1934)	= *aspera*
pricei	*Chingia* Holtt. (1974)	*Chingia* Holtt. (1974)
prionodes	*Polypodium* Wright (1906)	= *pauciflorum*
prismaticum	*Nephrodium* Desv. (1927)	*Pneumatopteris* Holtt. (1973)
procera	*Phegopteris* Mann (1868)	*Pseudophegopteris* Holtt. (1969)
procerum	*Nephrodium* Bak. (1874)	= *elatum* Boj.
procurrens	*Aspidium* Mett. (1864)	= *parasiticum* L.
producens	*Aspidium* Fée (1865)	= *Amauropelta*
productum	*Aspidium* Kaulf. (1824)	*Sphaerostephanos* Holtt. (1975)
prolatipedis	*Thelypteris* Lellinger (1977)	= *Amauropelta*
prolifera	*Hemionitis* Retz. (1791)	*Ampelopteris* Copel. (1947)
prolixum	*Aspidium* Willd. (1810)	*Christella* Holtt. (1973)
prominulum	*Aspidium* Christ (1896)	= *Thelypteris* s.l.
propinquoides	*Hypopeltis* Bory (1833)	= *interrupta* Willd.
propinquum	*Nephrodium* R.Br. (1810)	= *interrupta* Willd.
propria	*Dryopteris* v.A.v.R. (1914)	*Coryphopteris* Holtt. (1976)
protecta	*Dryopteris* Copel. (1942)	= *archboldii*
proximus	*Cyclosorus* Ching (1964)	= *Christella*
pseudoafricana	*Dryopteris* Makino & Ogata (1927)	= *omeiense*
pseudoamboinensis	*Dryopteris* Rosenst. (1917)	= *subpubescens* Bl.
pseudoarbuscula	*Dryopteris* v.A.v.R. (1916)	= *acrostichoides* Desv.
pseudoarfakiana	*Phegopteris* Hosok. (1938)	= *perglandulifera*
pseudoaspidoides	*Thelypteris* L. Gómez (1982)	= *Amauropelta*
pseudocalcarata	*Dryopteris* C. Chr. (1934)	= *sericea*
pseudocuspidata	*Dryopteris* Christ (1911)	= *penangianum*
pseudoferox	*Chingia* Holtt. (1974)	= *sakayensis*
pseudogueinziana	*Dryopteris* Bonap. (1913)	= *Christella*
pseudohirsuta	*Dryopteris* Rosenst. (1917)	= *productum* Kaulf.
x pseudoliukiuensis	*Thelypteris* Serizawa (1981)	= *Pronephrium*
pseudomegaphylla	*Dryopteris* v.A.v.R. (1917)	*Sphaerostephanos* Holtt. (1982)
pseudomontanum	*Aspidium* Hieron. (1896)	= *rivularioides*
pseudoparasitica	*Dryopteris* v.A.v.R. (1924)	= ? *Dryopteris*
pseudopatens	*Nephrodium* Jenm. (1896)	= *patens* Sw.
pseudoreptans	*Dryopteris* C. Chr. (1906)	= *debile* Bak.
pseudosancta	*Dryopteris* C. Chr. (1909)	= *Amauropelta*
pseudostenobasis	*Dryopteris* Copel. (1929)	*Amphineuron* Holtt. (1982)
pseudothelypteris	*Nephrodium* Rosenst. (1904)	= *Amauropelta*
psilophylla	*Pneumatopteris* Holtt. (1982)	*Pneumatopteris* Holtt. (1982)
ptarmicum	*Aspidium* Mett. (1858)	= *Amauropelta*
ptarmiciformis	*Dryopteris* Rosenst. (1913)	= *Amauropelta*
pterifolium	*Aspidium* Kuhn (1869)	= *Amauropelta*
pteroideum	*Polypodium* Klotzsch (1847)	= *Amauropelta*

Epithet	Genus in which Originally Described	Present or Proposed Disposition
pteroidea	Dryopteris C. Chr. (1906)	= terminans
pteroides	Aspidium Sw. (1801)	= interrupta Willd.
pteroides	Aspidium Mett. (1864)	= opulentum
pteroides	Nephrodium J. Sm. (1857)	= terminans
pterospora	Dryopteris v.A.v.R. (1920)	Sphaerostephanos Holtt. (1982)
puberulum	Aspidium Fée (1865)	Christella Löve & Löve (1977)
pubescens	Nephrodium Don (1825)	= cana
pubescens	Nephrodium Brack. (1854)	= heterocarpon
pubirachis	Nephrodium Bak. (1876)	Coryphopteris Holtt. (1976)
pulchrum	Aspidium Willd. (1810)	Pseudocyclosorus Holtt. (1974)
pullei	Plesioneuron Holtt. (1975)	Plesioneuron Holtt. (1975)
pullenii	Sphaerostephanos Holtt. (1982)	Sphaerostephanos Holtt. (1982)
punctatum	Nephrodium Bedd. (1866)	= opulentum
var. henryi	Hypolepis Christ (1905)	= torresianum var. calvatum
punctatus	Sphaerostephanos Holtt. (1975)	= productum Kaulf.
purdomii	Dryopteris C. Chr. (1913)	= levingei
purusense	Aspidium Christ (1906)	= hispidulum Dcne.
pusillum	Aspidium Mett. (1864)	= Amauropelta
pustulosus	Cyclosorus Copel. (1952)	= gymnopteridifrons
pycnosora	Dryopteris C. Chr. (1943)	Sphaerostephanos Holtt. (1977)
pygmaeus	Cyclosorus Ching & Zhang (1983)	= Christella
pyramidata	Goniopteris Fée (1866)	Goniopteris Fée (1866)
pyrrhorhachis	Polypodium Kze (1851)	Pseudophegopteris Ching (1963) & = paludosum Bl.
quadrangulare	Nephrodium Fée (1852)	= hilsenbergii
quadriaurita	Dryopteris Christ (1907)	= savaiense
quadriquetra	Dryopteris v.A.v.R. (1924)	Plesioneuron Holtt. (1975)
quaylei	Dryopteris E. Brown (1931)	Coryphopteris Holtt. (1976)
quelpaertensis	Dryopteris Christ (1910)	Oreopteris Holub (1969)
raddii	Dryopteris Rosenst. (1915)	= Amauropelta
raiateana	Coryphopteris Holtt. (1977)	Coryphopteris Holtt. (1977)
rammelooi	Macrothelypteris Pic. Ser. (1983)	= Pseudophegopteris
ramosii	Dryopteris Christ (1907)	Pronephrium Holtt. (1982)
rampans	Nephrodium Bak. (1889)	= penangianum
randallii	Thelypteris Maxon & Morton (1963)	= Amauropelta
rechingeri	Parathelypteris Holtt. (1977)	Parathelypteris Holtt. (1977)
reconditus	Sphaerostephanos Holtt. (1982)	Sphaerostephanos Holtt. (1982)
rectangulare	Polypodium Zoll. (1854)	Pseudophegopteris Holtt. (1969)
recumbens	Dryopteris Rosenst. (1906)	= Amauropelta

Epithet	Genus in which Originally Described	Present or Proposed Disposition
reducta	Dryopteris C. Chr. (1937)	= ekmanii
reducta	Thelypteris Small (1938)	= Amauropelta
reederi	Cyclosorus Copel. (1953)	= acrostichoides Desv.
reflexa	Dryopteris Ching (1931)	= erubescens Hook.
refractum	Polypodium Kze. (1850)	Goniopteris J. Sm. (1857)
regis	Dryopteris Copel. (1942)	Pneumatopteris Holtt. (1973)
regnelliana	Dryopteris C. Chr. (1907)	= Amauropelta
reineckei	Dryopteris C. Chr. (1943)	Sphaerostephanos Holtt. (1977)
reinwardtianum	Aspidium Kunze ex Mett. (1848)	= calcaratum Bl.
remotum	Nephrodium Heward (1842)	= dentatum
remotipinna	Dryopteris Bonap. (1917)	Pneumatopteris Holtt. (1973)
repanda	Goniopteris Fée (1852)	Pronephrium Holtt. (1972)
repandula	Dryopteris v.A.v.R. (1924)	= hispidulum Dcne.
repens	Nephrodium Hope (1899)	= canum
repentula	Dryopteris Christ (1909)	= glanduligerum
reptans	Polypodium J. F. Gmel. (1791)	Goniopteris Presl (1836)
resiliens	Dryopteris Maxon (1938)	= Goniopteris
resiniferum	Aspidium Kaulf. (1924)	= interrupta Willd.
resiniferum	Polypodium Desv. (1811)	Amauropelta Pic. Ser. (1977)
resinosofoetidum	Nephrodium Hook. (1862)	= cheilanthoides Kze
reticulatum	Polypodium L. (1759)	Meniscium Sw. (1803)
retrosum	Nephrodium Sod. (1883)	= Amauropelta
retusum	Polypodium Sw. (1817)	= raddii
rheophyta	Thelypteris Proctor (1985)	= Amauropelta
rhombea	Dryopteris Christ (1907)	Pronephrium Holtt. (1972)
richardsii	Nephrodium Bak. (1874)	Sphaerostephanos Holtt. (1977)
var. multifida	Lastrea T. Moore (1881)	= harveyi var. connivens
ridleyi	Lastrea Bedd. (1909)	= gymnopodum
riedleanum	Polypodium Mett. (1858)	= acrostichoides Desv.
rigescens	Nephrodium Sod. (1893)	= Amauropelta
rigida	Goniopteris Ridl. (1916)	Sphaerostephanos Holtt. (1982)
rigidifolia	Dryopteris v.A.v.R. (1924)	= badia
rigidilobum	Plesioneuron Holtt. (1975)	Plesioneuron Holtt. (1975)
rigidulum	Aspidium Mett. (1869)	= Amauropelta
rigidus	Cyclosorus Copel. (1952)	= productum Kaulf.
rimbachii	Dryopteris Rosenst. (1909)	= Amauropelta
riograndense	Polypodium Lindm. (1903)	Goniopteris Ching (1940)
riopardensis	Dryopteris Rosenst. (1906)	= Amauropelta
rioverdensis	Dryopteris C. Chr. (1907)	= Amauropelta
riparia	Dryopteris Copel. (1942)	= acrostichoides Desv.
riparium	Aspidium Bory (1810)	= tomentosum
riparium	Aspidium Moritz (1854)	= patens Sw.
rivoirei	Oochlamys Fée (1852)	= Amauropelta
rivulariformis	Dryopteris Rosenst. (1909)	= Amauropelta
rivularioides	Aspidium Fée (1869)	= Amauropelta
rivulorum	Polypodium Raddi (1825)	Amauropelta Pic. Ser. (1977)
robinsonii	Lastrea Ridl. (1920)	= gymnopodum

36

Epithet	Genus in which Originally Described	Present or Proposed Disposition
rodigasianum	Nephrodium Moore (1882)	Pneumatopteris Holtt. (1973)
roemeriana	Dryopteris Rosenst. (1912)	Sphaerostephanos Holtt. (1982)
rojasii	Dryopteris Christ (1909)	= Amauropelta
rolandii	Dryopteris C. Chr. (1913)	= Goniopteris
roraimense	Polypodium Bak. (1886)	= Amauropelta
rosei	Dryopteris Maxon (1915)	= Amauropelta
rosenburgii	Dryopteris C. Chr. (1934)	= arfakianum
rosenstockii	Dryopteris C. Chr. (1907)	= Amauropelta
rostrata	Goniopteris Fée (1866)	= glandulosum Desv.
rotumaensis	Cyclosorus H. St. John (1954)	= rodigasianum
royenii	Plesioneuron Holtt. (1975)	Plesioneuron Holtt. (1975)
rubicunda	Phegopteris v.A.v.R. (1920)	Pronephrium Holtt. (1972)
ssp. sulawesiense	Pronephrium Iwats. (1977)	Pronephrium Holtt. (1982)
rubidum	Polypodium Hook. (1863)	Pronephrium Holtt. (1972)
rubiformis	Dryopteris Robinson (1912)	= procerum Mann
rubra	Dryopteris Ching (1931)	= lakhimpurensis
rubrinervis	Phegopteris Mett. (1869)	Pronephrium Holtt. (1972)
rude	Polypodium Kze (1839)	Amauropelta Pic. Ser. (1977)
rudiformis	Dryopteris Copel. (1941)	= Amauropelta
rudis	Goniopteris Ridl. (1916)	Sphaerostephanos Holtt. (1982)
rufescens	Mesophlebion Holtt. (1975)	Mesophlebion Holtt. (1975)
rufum	Polypodium Poir. (1804)	= Amauropelta
rufopilosa	Dryopteris Brause (1920)	= rudis
rufostramineum	Aspidium Christ (1905)	Glaphyropteridopsis Ching (1963)
ruizianum	Polypodium Klotzsch (1847)	= Amauropelta
rupestre	Leptogramma Klotzsch (1847)	= Amauropelta
rupicola	Dryopteris C. Chr. (1917)	= Amauropelta
rupicola	Dryopteris Hosok. (1936)	Christella Holtt. (1977)
rupi-insularis	Thelypteris Fosberg (1980)	= rupicola Hosok.
rurutensis	Dryopteris Copel. (1938)	= costata Brack.
rusbyi	Dryopteris C. Chr. (1909)	= Amauropelta
rustica	Phegopteris Fée (1866)	= Amauropelta
rutteniana	Phegopteris v.A.v.R. (1918)	= beccarianum (Meniscium Ces.)
sagittatum	Polypodium Sw. (1788)	Goniopteris Pic. Ser. (1977)
sagittifolium	Aspidium Bl. (1828)	Sphaerostephanos Holtt. (1982)
sagittifolioides	Cyclosorus Copel. (1952)	= latebrosum
sakayense	Nephrodium Zeill. (1885)	Chingia Holtt. (1981)
salazica	Parathelypteris Holtt. (1974)	Amauropelta Holtt. (1976)
salicifolium	Meniscium Hook. (1854)	Pronephrium Holtt. (1972)
salzmannii	Meniscium Fée (1852)	Meniscium Fée (1852)
samarensis	Cyclosorus Copel. (1952)	Pronephrium Holtt. (1972)
sambasensis	Chingia Holtt. (1974)	Chingia Holtt. (1974)
sambiranensis	Dryopteris C. Chr. (1932)	= remotipinna
sampsonii	Polypodium Bak. (1891)	= megacuspe

Epithet	Genus in which Originally Described	Present or Proposed Disposition
sanctum	Acrostichum L. (1759)	Amauropelta Pic. Ser. (1977)
sanctiformis	Dryopteris C. Chr. (1913)	Amauropelta Löve & Löve (1977)
sanctoides	Aspidium Fée (1852)	= Amauropelta
sandsii	Plesioneuron Holtt. (1982)	Plesioneuron Holtt. (1982)
sandwicensis	Stegnogramma Brack. (1854)	Pneumatopteris Holtt. (1973)
sangnellii	Nephrodium Bak. (1891)	= prolixum Willd.
santae-catharinae	Dryopteris Rosenst. (1906)	= Amauropelta
santomasii	Sphaerostephanos Holtt. (1975)	Sphaerostephanos Holtt. (1975)
sarasinorum	Sphaerostephanos Holtt. (1982)	Sphaerostephanos Holtt. (1982)
savaiense	Nephrodium Bak. (1891)	Plesioneuron Holtt. (1975)
saxatilis	Thelypteris R.P. St. John (1939)	= Christella
saxicola	Polypodium Sw. (1817)	= Amauropelta
scaberulus	Cyclosorus Ching (1938)	Christella Holtt. (1976)
scabriuscula	Lastrea Presl (1836)	= patens Sw. var.
scabrum	Polypodium Presl (1822)	Goniopteris Brade (1972)
scabrum	Polypodium Roxb. (1844)	= ferox
scalare	Aspidium Christ (1906)	Amauropelta Löve & Löve (1977)
scallanii	Aspidium Christ (1901)	Stegnogramma Iwats. (1963)
scalpturata	Phegopteris Fée (1852)	= heteropterum
scalpturoides	Phegopteris Fée (1866)	= Amauropelta
scandens	Sphaerostephanos Holtt. (1977)	Sphaerostephanos Holtt. (1977)
scariosa	Dryopteris Rosenst. (1906)	= Amauropelta
schaffneri	Nephrodium Fée (1857)	= Goniopteris
schippii	Dryopteris Weath. (1935)	= Goniopteris
schizophylla	Dryopteris v.A.v.R. (1924)	= tuberculatum
schizotis	Nephrodium Hook. (1862)	= patens Sw. var. scabriuscula
schlechteri	Dryopteris Brause (1912)	= tuberculatum
schultzei	Dryopteris Brause (1912)	= keysseriana
schwackeana	Dryopteris Christ (1913)	Goniopteris Brade (1972)
schwenkii	Aspidium Bl. (ined.)	= terminans
sclerophyllum	Aspidium Spreng. (1827)	Goniopteris Wherry (1964)
scolopendrioides	Polypodium L. (1753)	Goniopteris Presl (1836)
scopulorum	Pronephrium Holtt. (1982)	Pronephrium Holtt. (1982)
seemännii	Coryphopteris Holtt. (1976)	Coryphopteris Holtt. (1976)
seemannii	Nephrodium Bak. (1891)	= globulifera
seemannii	Phegopteris J. Sm. (1854)	= Steiropteris
sellensis	Dryopteris C. Chr. (1937)	= Amauropelta
semicordatus	Sphaerostephanos Holtt. (1982)	Sphaerostephanos Holtt. (1982)
semihastatum	Aspidium Kze (1834)	= Thelypteris s.l.
semilunatum	Nephrodium Sod. (1883)	= Amauropelta
semimetralis	Sphaerostephanos Holtt. (1982)	Sphaerostephanos Holtt. (1982)
semisagittatum	Polypodium Roxb. (1844)	Christella Holtt. (1976)
sepikensis	Dryopteris Brause (1920)	= arfakianum

Epithet	Genus in which Originally Described	Present or Proposed Disposition
septemjuga	*Dryopteris* C. Chr. (1937)	= *Goniopteris*
septempedalis	*Dryopteris* Alston (1940)	*Plesioneuron* Holtt. (1975)
sericea	*Lastrea* Bedd. (1869)	*Trigonospora* Holtt. (1974)
serra	*Polypodium* Sw. (1788)	*Christella* Holtt. (1976)
serrata	*Tectaria* Cav. (1802)	= *unitum* (*Polypodium* L.)
serratum	*Meniscium* Cav. (1803)	*Meniscium* Cav. (1803)
serratus	*Cyclosorus* Copel. (1952)	= *productum* Kaulf.
serrulata	*Thelypteris* Ching (1936)	*Parathelypteris* Ching (1963)
serrulatum	*Polypodium* Sw. (1801)	*Goniopteris* J. Sm. (1857)
sessilifolium	*Polypodium* Hook. (1862)	= *cruciatum*
sessilipinna	*Dryopteris* Copel. (1911)	*Sphaerostephanos* Holtt. (1975)
setigera	*Cheilanthes* Bl. (1828)	*Macrothelypteris* Ching (1963) & = *torresianum*
var. *calvatum*	*Nephrodium* Bak. (1875)	= *torresianum* var.
var. *pallida*	*Dryopteris* v.A.v.R. (1908)	= *torresianum*
setosa	*Lastrea* Bedd. (1868)	= *Acystopteris tenuisecta*
setosula	*Thelypteris* Reed (1968)	= *angustifolium* Presl
setulosa	*Thelypteris* A. R. Smith (1980)	= *Steiropteris*
sevillana	*Thelypteris* Reed (1968)	= *glaber*
sewellii	*Nephrodium* Bak. (1876)	= *bergianum*
shaferi	*Dryopteris* Maxon & C. Chr. (1914)	= *Amauropelta*
sharpeana	*Polypodium* Bak. (1880)	= *aubertii*
sheringii	*Nephrodium* Jenm. (1879)	= *balbisii*
siambonense	*Aspidium* Hieron. (1897)	= *Amauropelta*
siamensis	*Thelypteris* Tagawa & Iwats. (1967)	*Christella* Holtt. (1976)
sibelana	*Pneumatopteris* Holtt. (1973)	*Pneumatopteris* Holtt. (1973)
sieberianum	*Polypodium* Kaulf. (1827)	= *heteropterum*
silvatica	*Goniopteris* Pappe & Rawson (1858)	= *unita* Kze
silviensis	*Dryopteris* Hieron. (1907)	= *Amauropelta*
simillima	*Dryopteris* C. Chr. (1906)	*Pronephrium* Holtt. (1972)
simozawae	*Thelypteris* Tagawa (1937)	= *hirsutipes*
simplex	*Meniscium* Hook. (1842)	*Pronephrium* Holtt. (1972)
simplicifolium	*Aspidium* J. Sm. ex Hook. (1854)	*Sphaerostephanos* Holtt. (1975)
var. *vitiensis*	*Goniopteris* Carr. (1893)	= *beccarianum* (*Meniscium* Ces.)
simplicifrons	*Dryopteris* C. Chr. (1906)	= *Meniscium*
simplicissimum	*Aspidium* Christ (1904)	= *Amauropelta*
simulans	*Nephrodium* Bak. (1888)	= *simillima*
simulans	*Nephrodium* Bak. (1890)	= *fasciculatum*
simulans	*Thelypteris* Ching (1936)	= *auriculata*
simulatum	*Aspidium* Davenp. (1894)	*Parathelypteris* Holtt. (1976)
sinensis	*Craspedosorus* Ching & W. M. Chu (1978)	*Craspedosorus* Ching & W. M. Chu (1978)
singalanense	*Nephrodium* Bak. (1880)	*Metathelypteris* Ching (1963)
sinica	*Dryopteris* Christ (1909)	= *acuminatum* Houtt.

Epithet	Genus in which Originally Described	Present or Proposed Disposition
sinodentatus	Cyclosorus Ching & Liu (1984)	= Christella
sintense	Aspidium Krug (1897)	= sclerophyllum var.
skinneri	Aspidium Hook. (1854)	= Goniopteris
smithianum	Nephrodium Presl (1851)	acrostichoides Desv.
smithii	Phegopteris v.A.v.R. (1908)	= beddomei
sodiroi	Thelypteris Reed (1968)	= nemorale Sod.
sogerensis	Dryopteris Gepp (1923)	Pneumatopteris Holtt. (1973)
solsonicum	Pronephrium Holtt. (1974)	Pronephrium Holtt. (1974)
solutus	Sphaerostephanos Holtt. (1982)	Sphaerostephanos Holtt. (1982)
somai	Dryopteris Hayata (1915)	= Pseudophegopteris
sophoroides	Nephrodium Thunb. (1794)	= acuminatum Houtt.
sorbifolium	Meniscium Desrouss. (1797)	Meniscium Desrouss. (1797)
spenceri	Dryopteris Christ (1907)	Sphaerostephanos Holtt. (1975)
splendens	Glaphyropteridopsis Ching (1963)	Glaphyropteridopsis Ching (1963)
sprengelii	Aspidium Kaulf. (1823)	= Amauropelta
sprucei	Nephrodium Bak. (1867)	= pachyrachis var.
squamaestipes	Polypodium Clarke (1880)	Cyclogramma Tagawa (1938)
squamatellus	Sphaerostephanos Holtt. (1982)	Sphaerostephanos Holtt. (1982)
squamigerum	Aspidium Schlecht. (1825)	= confluens Thunb.
squamipes	Dryopteris Copel. (1935)	Coryphopteris Holtt. (1976)
squamulosum	Nephrodium Hook. f. (1855)	= confluens Thunb.
srilankensis	Thelypteris Panigrahi (1975)	= zeylanicum Fée
standleyi	Dryopteris Maxon & Morton (1938)	Meniscium Pic. Ser. (1968)
stegnogramma	Gymnogramme Bl. (1828)	= aspidioides Bl.
stegnogrammoides	Polypodium Bak. (1867)	= sandwicensis
stellatopilosa	Dryopteris Brause (1920)	= Diplazium
stenobasis	Dryopteris C. Chr. (1906)	= attenuatum Kunze ex Mett.
stenodonta	Cyclosorus Copel. (1952)	Sphaerostephanos Holtt. (1975)
stenolepis	Polypodium Bak. (1898)	= Dryopteris
stenophylla	Cheilanthes Kze (1848)	= setigera
stenophylla	Dryopteris Rosenst. (1909)	= rivulariformis
stenophyllum	Meniscium Bak. (1891)	= exsculptum
stenophyllum	Nephrodium Sod. (1883)	= Amauropelta
stenopodum	Pronephrium P. Chandra (1971)	Pronephrium P. Chandra (1971)
stenura	Plesioneuron Holtt. (1982)	Plesioneuron Holtt. (1982)
stereophylla	Dryopteris v.A.v.R. (1924)	Coryphopteris Holtt. (1976)
steyermarkii	Thelypteris A. R. Smith (1983)	= Amauropelta
stierii	Gymnogramme Rosenst. (1905)	= Amauropelta
stipellatum	Aspidium Bl. (1828)	Sphaerostephanos Holtt. (1982)
var. obtusata	Dryopteris v.A.v.R. (1920)	= batacorum

Epithet	Genus in which Originally Described	Present or Proposed Disposition
stipulaceum	Aspidium Mett. (1858)	= heteropterum
stipulare	Aspidium Willd. (1810)	= patens Sw.
stokesii	Dryopteris E. Brown (1931)	Pneumatopteris Holtt. (1973)
stolzeana	Thelypteris A. R. Smith (1976)	= Goniopteris
straminea	Phegopteris Fée (1852)	= cruciatum
stramineum	Nephrodium Sod. (1883)	= Amauropelta
stramineum	Polypodium Bak. (1867)	= Goniopteris
stresemannii	Sphaerostephanos Holtt. (1982)	Sphaerostephanos Holtt. (1982)
striatum	Aspidium Schum. (1829)	Cyclosorus Ching (1941)
strigifera	Dryopteris Hieron. (1907)	= Amauropelta
strigosa	Goniopteris Fée (1866)	= scolopendrioides
strigosum	Aspidium Willd. (1810)	= Amauropelta
strigosissima	Dryopteris Copel. (1942)	= hispiduliformis var. brassii
struthiopteroides	Dryopteris C. Chr. (1909)	= Amauropelta
stuebelii	Dryopteris Hieron. (1907)	= thomsonii
subalpina	Dryopteris v.A.v.R. (1922)	Sphaerostephanos Holtt. (1982)
subandina	Dryopteris Rosenst. (1913)	= Amauropelta
subappendiculata	Dryopteris Copel. (1942)	Pneumatopteris Holtt. (1973)
subaridus	Cyclosorus Tagawa (1938)	Christella C. M. Kuo (1975)
subattenuata	Dryopteris Rosenst. (1912)	Amphineuron Holtt. (1977)
subaurita	Dryopteris Tagawa (1932)	Pseudophegopteris Ching (1963)
subbipinnata	Coryphopteris Holtt. (1976)	Coryphopteris Holtt. (1976)
subconformis	Dryopteris C. Chr. (1933)	= amboinense Willd.
subcordatus	Sphaerostephanos Holtt. (1982)	Sphaerostephanos Holtt. (1982)
subcuneatum	Nephrodium Bak. (1870)	= Thelypteris s.l.
subcuspidata	Dryopteris Rosenst. (1917)	= penangianum
subdecussatum	Aspidium Christ (1904)	= Amauropelta
subdentata	Christella Holtt. (1976)	Christella Holtt. (1976)
subdimorpha	Dryopteris Copel. (1942)	= crassifolium Bl.
subdimorphus	Cyclosorus Copel. (1952)	= granulosum
subelatum	Nephrodium Bak. (1906)	Christella Holtt. (1976)
subfalcinella	Dryopteris v.A.v.R. (1920)	= norrisii
subfuscum	Nephrodium Bak. (1867)	= Steiropteris
subglabrum	Plesioneuron Holtt. (1975)	Plesioneuron Holtt. (1975)
subglandulifera	Thelypteris Ching (1938)	= pectiniformis
subglandulosum	Nephrodium Bak. (1867)	= cruciatum
subhispidula	Dryopteris Rosenst. (1915)	= taiwanensis
subimmersa	Thelypteris Ching (1936)	= immersum
subintegrum	Polypodium Bak. (1877)	= Thelypteris s.l.
subjunctum	Nephrodium Bak. (1891)	Christella Holtt. (1977)
sublaevifrons	Dryopteris Tagawa (1936)	= truncatum Poir.
sublaxa	Dryopteris Hayata (1914)	= gracilescens Bl.
submollis	Dryopteris v.A.v.R. (1920)	= dentatum
subnigra	Dryopteris Brause (1920)	Coryphopteris Holtt. (1976)
subochthodes	Thelypteris Ching (1936)	= esquirolii
suboppositus	Sphaerostephanos Holtt. (1982)	Sphaerostephanos Holtt. (1982)

Epithet	Genus in which Originally Described	Present or Proposed Disposition
subpectinata	Dryopteris Copel. (1932)	Sphaerostephanos Holtt. (1977)
subpennigera	Dryopteris C. Chr. (1932)	Pneumatopteris Holtt. (1973)
subpubescens	Aspidium Bl. (1828)	Christella Holtt. (1976)
subsagittata	Dryopteris C. Chr. (1937)	= Goniopteris
subsimile	Polypodium Colenso (1888)	= pennigerum Forst.
subsimilis	Gymnogramme Hook. (1864)	= Thelypteris s.l.
subspinosa	Dryopteris C. Chr. (1906)	= glandulifera
subterminale	Plesioneuron Holtt. (1975)	Plesioneuron Holtt. (1975)
subtetragonum	Polypodium Link (1833)	Goniopteris Presl (1836)
subthelypteris	Dryopteris C. Chr. (1906)	= flexile
subtilis	Thelypteris A. R. Smith (1983)	= Amauropelta
subtruncatum	Polypodium Bory (1833)	Sphaerostephanos Holtt. (1971)
subulifolia	Dryopteris v.A.v.R. (1918)	Sphaerostephanos Holtt. (1982)
subvillosa	Thelypteris Ching (1936)	= auriculata
x subviridifrons	Thelypteris Serizawa (1981)	= Macrothelypteris
subviscosa	Dryopteris v.A.v.R. (1915)	= gymnopodum
sudesticus	Sphaerostephanos Holtt. (1982)	Sphaerostephanos Holtt. (1982)
sulphurea	Dryopteris E. Brown (1931)	= opulentum
sumatrana	Dryopteris v.A.v.R. (1908)	= subpubescens Bl.
sumatrana	Pseudophegopteris Holtt. (1969)	Pseudophegopteris Holtt. (1969)
sumatrensis	Mesochlaena v.A.v.R. (1920)	= larutense
sumbawensis	Dryopteris C. Chr. (1934)	Pneumatopteris Holtt. (1973)
superba	Dryopteris Brause (1920)	Pneumatopteris Holtt. (1973)
superficialis	Dryopteris v.A.v.R. (1913)	= attenuatum Kunze ex Mett.
supinum	Nephrodium Sod. (1893)	= Amauropelta
supralineata	Dryopteris Rosenst. (1910)	= Steiropteris
supranitens	Dryopteris Christ (1910)	Amauropelta Löve & Löve (1977)
supraspinigera	Dryopteris Rosenst. (1915)	Chingia Holtt. (1974)
suprastrigosa	Dryopteris Rosenst. (1912)	= heterocarpon
supravillosa	Dryopteris C. Chr. (1934)	= obtusata v.A.v.R.
x tabaquitensis	Goniopteris Jermy & Walker (1985)	Goniopteris Jermy & Walker (1985)
tablanum	Aspidium Christ (1905)	= Amauropelta
tablaziensis	Dryopteris Christ (1907)	= Amauropelta
tahanensis	Coryphopteris Holtt. (1976)	Coryphopteris Holtt. (1976)
tahitensis	Plesioneuron Holtt. (1975)	Plesioneuron Holtt. (1975)
taiwanensis	Dryopteris C. Chr. (1906)	Sphaerostephanos Kuo (1975)
talamauensis	Mesochlaena v.A.v.R. (1918)	Sphaerostephanos C. Chr. (1934)
tamandarei	Dryopteris Rosenst. (1915)	= Amauropelta

Epithet	Genus in which Originally Described	Present or Proposed Disposition
tamiensis	Dryopteris Brause (1912)	= ? Sphaerostephanos
tandikatensis	Dryopteris v.A.v.R. (1913)	Sphaerostephanos Holtt. (1982)
tanggamensis	Coryphopteris Holtt. (1976)	Coryphopteris Holtt. (1976)
tannensis	Dryopteris C. Chr. (1906)	= glandulifera Brack.
taprobanica	Thelypteris Panigrahi (1976)	Christella Holtt. (1976)
tatei	Dryopteris Maxon & Morton (1938)	= Steiropteris
tectum	Nephrodium Bedd. (1892)	= parasiticum L.
telefominicus	Sphaerostephanos Holtt. (1982)	Sphaerostephanos Holtt. (1982)
tenebricum	Nephrodium Jenm. (1882)	= Thelypteris s.l.
tenella	Phegopteris Fée (1852)	= Amauropelta
tenera	Goniopteris Fée (1866)	= reptans var.
tenericaule	Polypodium Hook. (1857)	= torresianum
tenerior	Chingia Holtt. (1982)	Chingia Holtt. (1982)
tenerrimum	Aspidium Fée (1869)	= Amauropelta
tenggerensis	Pseudophegopteris Holtt. (1974)	Pseudophegopteris Holtt. (1974)
tenompokensis	Dryopteris C. Chr. (1934)	= peltata var.
tenuicula	Aspidium Fée (1852)	= Amauropelta
tephrophylla	Dryopteris Copel. (1929)	Sphaerostephanos Holtt. (1975)
terminans	Nephrodium Hook. (1862)	Amphineuron Holtt. (1973)
terrestris	Dryopteris Copel. (1942)	= confertus
tetragonum	Polypodium Sw. (1788)	= Amauropelta
teuscheri	Dryopteris v.A.v.R. (1908)	Mesophlebion Holtt. (1975)
thelypteris	Acrostichum L. (1753)	= palustris Schott.
var. squamigerum	Aspidium Schlecht. (1825)	= confluens Thunb.
var. squamulosum	Nephrodium Hook. (1862)	= confluens Thunb.
thelypteroides	Nephrodium Michx. (1803)	= palustris Schott. var. pubescens.
thomsonii	Polypodium Jenm. (1886)	Amauropelta Pic. Ser. (1977)
thwaitesii	Meniscium Hook. (1859)	Pronephrium Holtt. (1972)
thysanoides	Pronephrium Holtt. (1982)	Pronephrium Holtt. (1982)
tibangensis	Dryopteris C. Chr. (1937)	Sphaerostephanos Holtt. (1982)
tibetana	Pseudophegopteris Ching & Wu (1983)	Pseudophegopteris Ching & Wu (1983)
tibetica	Cyclogramma Ching & Wu (1983)	Cyclogramma Ching & Wu (1983)
tibetica	Metathelypteris Ching & Wu (1983)	Metathelypteris Ching & Wu (1983)
tibetica	Phegopteris Ching (1983)	Phegopteris Ching (1983)
tibeticus	Pseudocyclosorus Ching & Ling (1984)	Pseudocyclosorus Ching & Ling (1984)
tildeniae	Amphineuron Holtt. (1977)	Amphineuron Holtt. (1977)
timorensis	Christella Holtt. (1976)	Christella Holtt. (1976)
tobaica	Pneumatopteris Holtt. (1973)	Pneumatopteris Holtt. (1973)
todayensis	Dryopteris Christ (1907)	= latebrosum
toganetra	Thelypteris A. R. Smith (1973)	Goniopteris Löve & Löve (1977)

Epithet	Genus in which Originally Described	Present or Proposed Disposition
tomentosum	*Polypodium* Thouars (1804)	*Amauropelta* Holtt. (1977)
tonkinensis	*Dryopteris* C. Chr. (1934)	*Amphineuron* Holtt. (1977)
toppingii	*Dryopteris* Copel. (1917)	= *neotoppingii*
toppingii	*Mesochlaena* Copel. (1917)	= *polycarpon* Bl.
torresianum	*Polystichum* Gaud. (1824)	*Macrothelypteris* Ching (1963)
tottum	*Polypodium* Thunb. (1800)	*Cyclosorus* Pic. Ser. (1968)
tottum	*Polypodium* Willd. (1810)	= *pozoi*
var. *subcalcarata*	*Phegopteris* v.A.v.R. (1917)	*Stegnogramma* Holtt. (1982)
tottoides	*Leptogramma* Ito (1935)	*Stegnogramma* Iwats. (1963)
trachyphyllum	*Pronephrium* Holtt. (1982)	*Pronephrium* Holtt. (1982)
translucens	*Plesioneuron* Holtt. (1975)	*Plesioneuron* Holtt. (1975)
transversarium	*Nephrodium* Brack. (1854)	*Pneumatopteris* Holtt. (1973)
trelawniensis	*Thelypteris* Proctor (1981)	= *Amauropelta*
tremula	*Dryopteris* Christ (1910)	= *Thelypteris*
trichochlamys	*Sphaerostephanos* Holtt. (1982)	*Sphaerostephanos* Holtt. (1982)
trichodes	*Polypodium* Houlst. & Moore (1851)	= *torresianum*
trichodes	*Dryopteris* Rosenst. (1917)	= *polypodioides* Hook. & *torresianum*
trichophorum	*Aspidium* Fée (1866)	= *l'herminieri*
trichopoda	*Dryopteris* C. Chr. (1906)	*Mesophlebion* Holtt. (1975)
trimetralis	*Sphaerostephanos* Holtt. (1982)	*Sphaerostephanos* Holtt. (1982)
triphyllum	*Meniscium* Sw. (1801)	*Pronephrium* Holtt. (1972)
triste	*Polypodium* Kze (1834)	*Goniopteris* Brade (1972)
truncata	*Abacopteris* Fée (1852)	= *menisciicarpon*
truncatum	*Polypodium* Poir. (1804)	*Pneumatopteris* Holtt. (1973)
truncatum	*Polystichum* Gaud. (1827)	= *lawakii*
var. *celebicum*	*Aspidium* Miquel (1869)	= ? *auctipinna*
tsangii	*Cyclosorus* Ching (ined)	= *parasiticum* L.
tsaratananensis	*Dryopteris* C. Chr. (1941)	= *oppositiformis*
tuberculatum	*Nephrodium* Ces. (1877)	*Plesioneuron* Holtt. (1975)
tuberculifera	*Dryopteris* C. Chr. (1931)	= *tylodes*
tuerckheimii	*Nephrodium* Donn-Sm. (1887)	= *Christella*
turrialbae	*Dryopteris* Rosenst. (1925)	*Meniscium* Pic. Ser. (1968)
tylodes	*Aspidium* Kze (1851)	*Pseudocyclosorus* Ching (1963)
uaniensis	*Sphaerostephanos* Holtt. (1982)	*Sphaerostephanos* Holtt. (1982)
ugoensis	*Thelypteris* Reed (1968)	= *productum* Kaulf.
uliginosum	*Aspidium* Kze (1847)	= *cosmopolita, setigera* & *torresianum*
var. *elegans*	*Dryopteris* Koidz. (1924)	= *torresianum* var. *calvatum*
unca	*Thelypteris* R.P. St. John (1938)	= *Christella*
uncinata	*Thelypteris* A. R. Smith	= *Amauropelta*
underwoodiana	*Dryopteris* Maxon (1928)	= *Amauropelta*

Epithet	Genus in which Originally Described	Present or Proposed Disposition
uniauriculata	*Dryopteris* Copel. (1914)	*Sphaerostephanos* Holtt. (1982)
unidentata	*Lastrea* Bedd. (1892)	*Coryphopteris* Holtt. (1976)
unijugus	*Sphaerostephanos* Copel. (1936)	= *polycarpon* Bl.
unita	*Gymnogramme* Kze. (1844)	*Pneumatopteris* Holtt. (1973)
unitum	*Aspidium* Mett. (1864)	= *interrupta* Willd.
unitum	*Nephrodium* Hook. & Arn. (1833)	= *terminans*
unitum	*Polypodium* L. (1759)	*Sphaerostephanos* Holtt. (1974)
unitum	*Polypodium* Thunb. (1874)	= *acuminatum* Houtt.
uraiensis	*Dryopteris* Rosenst. (1915)	*Metathelypteris* Ching (1963)
urbanii	*Nephrodium* Sod. (1893)	= *Goniopteris*
urdanetensis	*Dryopteris* Copel. (1913)	*Sphaerostephanos* Holtt. (1982)
urens	*Chingia* Holtt. (1974)	*Chingia* Holtt. (1974)
urens	*Dryopteris* Rosenst. (1907)	= *conspersum*
urophylla	*Phegopteris* Mett. (1858)	= *repanda*
var. *nitida*	*Dryopteris* Holtt. (1934)	*Pronephrium* Holtt. (1972)
var. *novoguineensis*	*Dryopteris* Rosenst. (1912)	= *aspera*
var. *peraspera*	*Dryopteris* v.A.v.R. (1922)	= *menisciicarpon*
var. *teysmannii*	*Dryopteris* v.A.v.R. (1918)	= *euryphylla*
var. *uniseriale*	*Polypodium* Hook. (1863)	= *gymnopteridifrons*
usambarensis	*Pneumatopteris* Holtt. (1973)	*Pneumatopteris* Holtt. (1973)
usitatum	*Nephrodium* Jenm. (1879)	= *Goniopteris*
utanagensis	*Dryopteris* Hieron. (1907)	= *Amauropelta*
vaccanea	*Dryopteris* Bosco (1938)	= *Amauropelta*
valdepilosum	*Nephrodium* Bak. (1879)	*Steiropteris* Pic. Ser. (1973)
valida	*Dryopteris* Christ (1908)	*Sphaerostephanos* Holtt. (1979)
vandervekenii	*Metathelypteris* Pic. Ser. (1983)	*Metathelypteris* Pic. Ser. (1983)
vanheurckii	*Aspidium* Fourn. (1872)	= *Amauropelta*
varians	*Nephrodium* Fée (1866)	= *Stigmatopteris*
x *varievenulosa*	*Thelypteris* Viane (1985)	= *Christella* x *Pneumatopteris*
varievestita	*Dryopteris* C. Chr. (1937)	*Plesioneuron* Holtt. (1982)
vattuonei	*Dryopteris* Hicken (1924)	= *Amauropelta*
vaupelii	*Dryopteris* C. Chr. (1943)	*Pneumatopteris* Holtt. (1973)
veitchii	*Sphaerostephanos* Holtt. (1977)	*Sphaerostephanos* Holtt. (1977)
velutinum	*Polypodium* Sod. (1883)	= *decussatum* var.
venulosum	*Aspidium* Bl. (1828)	= *interrupta* Willd.
venulosum	*Nephrodium* Desv. (1827)	= *gongylodes* var.
venulosum	*Nephrodium* Hook. (1862)	*Pneumatopteris* Holtt. (1973)
venustum	*Aspidium* Heward (1838)	*Goniopteris* Pic. Ser. (1977)
verecunda	*Thelypteris* Proctor (1985)	= *Goniopteris*
verrucosa	*Lastrea* Presl (1851)	= *immersum*
verruculosa	*Dryopteris* v.A.v.R. (1915)	= *lineatum* Bl.
var. *sumatrana*	*Dryopteris* v.A.v.R. (1920)	= *menisciicarpon*

Epithet	Genus in which Originally Described	Present or Proposed Disposition
versicolor	Thelypteris Small (1938)	= Christella
versteeghii	Pneumatopteris Holtt. (1973)	Pneumatopteris Holtt. (1973)
vestigiata	Dryopteris Copel. (1942)	Sphaerostephanos Holtt. (1982)
vile	Aspidium Racib. (1898)	= setigera
villosa	Gymnogramme Link (1833)	= Stegnogramma
villosipes	Dryopteris Gepp (1917)	= badia
vinosicarpa	Dryopteris v.A.v.R. (1922)	= motleyanum
violascens	Aspidium Link (1833)	= dentatum
viridifrons	Thelypteris Tagawa (1936)	Macrothelypteris Ching (1963)
viridis	Cyclosorus Copel. (1952)	= glaber
viscosum	Nephrodium Bak. (1867)	Coryphopteris Holtt. (1971)
vitiensis	Coryphopteris Holtt. (1976)	Coryphopteris Holtt. (1976)
vivipara	Polypodium Raddi (1925)	Goniopteris Brade (1972)
vulcanicum	Nephrodium Bak. (1894)	= gracilescens Bl.
vulgaris	Phegopteris Mett. (1856)	= connectile
wagneri	Thelypteris Fosb. & Sachet (1972)	= terminans
wakefieldii	Nephrodium Bak. (1891)	= opulentum
walkeri	Pneumatopteris Holtt. (1973)	Pneumatopteris Holtt. (1973)
wallichii	Dryopteris Rosenst. (1917)	= squamaestipes Clarke
wangii	Cyclosorus Ching (1964)	= Christella
wantotense	Mesoneuron sensu Holtt. in Blumea 13 (1965)	= fulgens
wantotensis	Dryopteris Copel. (1942)	Plesioneuron Holtt. (1975)
warburgii	Aspidium Kuhn & Christ (1900)	Sphaerostephanos Holtt. (1977)
wariensis	Dryopteris Copel. (1911)	Plesioneuron Holtt. (1975)
warmingii	Dryopteris C. Chr. (1913)	Goniopteris Brade (1972)
wauensis	Sphaerostephanos Holtt. (1982)	Sphaerostephanos Holtt. (1982)
weberi	Dryopteris Copel. (1929)	= dichotrichoides
weberi	Cyclosorus Copel. (1952)	= productum Kaulf.
x wildemanii	Dryopteris Christ (1909)	= Christella x Pneumatopteris
wilfordii	Hemionitis Hook. (1859)	= griffithii Moore var.
williamsii	Dryopteris Copel. (1931)	Sphaerostephanos Holtt. (1975)
woitapensis	Sphaerostephanos Holtt. (1982)	Sphaerostephanos Holtt. (1982)
wollastonii	Phegopteris v.A.v.R. (1917)	= rigida Ridl.
womersleyi	Pronephrium Holtt. (1972)	Pronephrium Holtt. (1972)
woodlarkensis	Cyathea Copel. (1914)	Plesioneuron Holtt. (1975)
wrightii	Aspidium D. C. Eaton (1860)	Steiropteris Pic. Ser. (1973)
xiphioides	Dryopteris Christ (1907)	Pronephrium Holtt. (1972)
xylodes	Aspidium Kze (1851)	= tylodes
yakumontana	Dryopteris Masam. (1932)	= quelpaertensis var.

Index of Thelypteridaceae

Epithet	Genus in which Originally Described	Present or Proposed Disposition
yamawakii	*Cyclosorus* Ito (1952)	= *acuminatum* Houtt. var.
yandongensis	*Cyclosorus* Ching & Shing (1983)	= *Christella*
yigongensis	*Pseudophegopteris* Ching (1983)	*Pseudophegopteris* Ching (1983)
yunkweiensis	*Thelypteris* Ching (1936)	*Pseudophegopteris* Ching (1963)
yunnanense	*Aspidium* Christ (1898)	= *Kuniwatsukia*
zambesiacum	*Nephrodium* Bak. (1891)	= *pulchrum*
zamboangensis	*Thelypteris* Reed (1968)	= *christii*
zayuensis	*Pseudocyclosorus* Ching & Wu (1983)	*Pseudocyclosorus* Ching & Wu (1983)
zayuensis	*Pseudophegopteris* Ching & Wu (1983)	*Pseudophegopteris* Ching & Wu (1983)
zeylanica	*Thelypteris* Ching (1936)	*Trigonospora* Sledge (1981)
zeylanicum	*Nephrodium* Fée (1865)	*Christella* Holtt. (1974)
zippelii	*Dryopteris* Rosenst. (1917)	= *peltata*

SELECTED BIBLIOGRAPHY

Abbiatti, D. (1964). Estudios sobre Pteridófitas austramericas de los géneros *Thelypteris*, *Cyclosorus* y *Goniopteris*. Darwiniana 13: 537–567.

Brade, A.C. (1964). Filices novae Brasilienses VIII. Arq. Jard. Bot. Rio Janeiro, 18:25–34.

— (1972). O gênero *Dryopteris* (Pteridophyta) no Brasil e sua divisão taxonômica. Bradea 1(22): 191–261.

Brownsey, P. J. & Jermy, A.C. (1973). A fern collecting expedition to Crete. Brit. Fern Gaz. 10(6): 331–348.

Ching, R.C. (1963). A reclassification of the family Thelypteridaceae from the mainland of Asia. Acta Phytotax. Sin. 8(4): 289–335.

— (1964). Additional material for the Pteridophytioflora of Hainan. Acta Phytotax. Sin. 9(4): 345–373.

— (1978). The Chinese fern families and genera: systematic arrangement and historical origin. Acta Phytotax. Sin. 16(3): 1–19.

— (1978). The Chinese fern families and genera: systematic arrangement and historical origin (cont.). Acta Phytotax. Sin. 16(4): 16–37.

— & Wu, S.K. (1983) Pteridophyta, in Flora Xizangica Vol. 1 ed. Wu C.Y.

Copeland, E.B. (1960). Fern flora of the Philippines.

Gilli, A. (1978). Beiträge zur Flora von Papua-New Guinea. Ann. Naturhist. Mus. Wien 81: 19–29.

Gomez, L.D. (1976). Contribuciones a la pteridología centroamericana 1. Enumeratio Filicum Nicaraguensium. Brenesia 8: 41–56.

Holttum, R.E. (1969). Studies in the family Thelypteridaceae. The genera *Phegopteris*, *Pseudophegopteris* and *Macrothelypteris*. Blumea 17(1): 5–32.

— (1971). Studies in the family Thelypteridaceae III. A new system of genera in the Old World. Blumea 19(1): 17–52.

— (1972). Studies in the family Thelypteridaceae IV. The genus *Pronephrium* Presl. Blumea 20(1): 105–126.

— (1973). The identity of three types in the Willdenow herbarium. Amer. Fern. Journ. 63(3): 81–84.

— (1973). Studies in the family Thelypteridaceae V. The genus *Pneumatopteris* Nakai. Blumea 21(2): 293–325.

— (1973). Studies in the family Thelypteridaceae VI. *Haplodictyum* and *Nannothelypteris*. Kalikasan, Philipp. Journ. Biol. 2: 58–68.

— (1974). Studies in the family Thelypteridaceae VII. The genus *Chingia*. Kalikasan, Philipp. Journ. Biol. 3: 13–28.

— (1974). The genus *Trigonospora* (Thelypteridaceae) in Malesia. Reinwardtia 8(4): 503–507.

— (1974). Additions to the Fern Flora of Java. Reinwardtia 8(4): 499–501.

— (1974). Thelypteridaceae from Africa and adjacent Islands. Journ. S. Afr. Bot. 40(2): 123–168.

— (1975). Studies in the family Thelypteridaceae VIII. The genera *Mesophlebion* and *Plesioneuron*. Blumea 22(2): 233–250.

— (1975). Studies in the family Thelypteridaceae IX. The genus *Sphaerostephanos* in the Philippines. Kalikasan, Philipp. Journ. Biol. 4: 47–68.

— (1976). Two new combinations in the genus *Amauropelta* (Thelypteridaceae). Kew Bull. 30(4): 607–608.

— (1976). Studies in the family Thelypteridaceae X. The genus *Coryphopteris*. Blumea 23(1): 18–47.

— (1976). Studies in the family Thelypteridaceae XI. The genus *Christella* Léveillé sect. *Christella*. Kew Bull. 31(2): 293–339.

— (1976). New records of Thelypteridaceae from the Philippines. Kalikasan, Philipp. Journ. Biol. 5: 109–120.

— (1976). Some new names in Thelypteridaceae with comments on cytological reports relating to the family. Webbia 30(1): 191–195.

— (1977). Studies in the family Thelypteridaceae XII. The genus *Amphineuron* Holttum. Blumea 23(2): 205–218.

— (1977). The family Thelypteridaceae in the Pacific and Australasia. Allertonia 1(3): 169–234.

— (1979). *Sphaerostephanos* (Thelypteridaceae) in Asia, excluding Malesia. Kew Bull. 34(2): 221–232.

— (1982). Thelypteridaceae. Flora Malesiana Series 2, Vol. 1(5): 331–599.

Holttum, R.E. & Chandra, P. (1971). New species of Thelypteridaceae from India, Ceylon and Burma. Kew Bull. 26(1): 79–82.

Holttum, R.E. & Grimes, J.W. (1979). The genus *Pseudocyclosorus* Ching (Thelypteridaceae). Kew Bull. 34(3): 499–516.

Holub, J. (1969). *Oreopteris*, a new genus of the family Thelypteridaceae. Folia Geobot. Phytotax. 4: 33–53.

Iwatsuki, K. (1963). Taxonomic studies of Pteridophyta VII. A revision of the genus *Stegnogramma* emend. Acta Phytotax. Geobot. 19(4–6): 112–126.

— (1964). An American species of *Stegnogramma*. Amer. Fern. Journ. 54(3): 141–143.

Kramer, K.U. (1969). Two new species of ferns from Suriname. Acta Bot. Neerl. 18(1): 138–142.

Kuo, C.M. (1975). Flora of Taiwan 1.

Lellinger, D.B. (1977). Nomenclatural notes on some ferns of Costa Rica, Panama and Colombia. Amer. Fern. Journ. 67(2): 58–60.

Löve, A. & Löve, D. (1977). New combinations in ferns. Taxon 26(2–3): 324–326.

Morton, C.V. (1971). The proper disposition of *Meniscium macrophyllum* Kunze. Amer. Fern. Journ. 61(1): 17–20.

— (1973). Studies of fern types, II. Contr. U.S. Nat. Herb. 38(6): 215–281.

Nayar, B.K. & Kaur, S. (1974). Companion to Beddome's handbook to the ferns of British India.

Pichi Sermolli R.E.G. (1968). Fragmenta pteriodologiae I. Webbia 23(1): 159–207.

— (1973). Fragmenta pteriodologiae IV. Webbia 28(2): 445–477.

— (1977). Fragmenta pteridologiae VI. Webbia 31(1): 237–259.

Proctor, G.R. (1961). Notes on Lesser Antilles ferns. Rhodora 63: 31–35.

— (1966). Notes on Lesser Antilles ferns II. Rhodora 68: 464–469.

— (1985). New species of Thelypteris from Puerto Rico. Amer. Fern Jour. 75: 56–70.

Reed, C.F. (1968). Index thelypteridis. Phytologia 17(4): 249–466.

— (1969). Index thelypteridis. Suppl. 1. Phytologia 17(7): 465–6.

Rodriguez, R.R. (1972). In Rodriguez & Duek. Lista preliminar de las especies de Pteridophyta en Chile continental e insular. Bol. Soc. Biol. Concepcion 45: 129–174.

Sledge, W.A. (1981). The Thelypteridaceae of Ceylon. Bull. Brit. Mus. (Nat. Hist.) Bot., 8(1):1–54.

Smith, A.R. (1971). Systematics of the neotropical species of *Thelypteris* section *Cyclosorus*. Univ. Calif. Publ. Bot. 59: 1–143.

— (1973). The Mexican species of *Thelypteris* subgenera *Amauropelta* and *Goniopteris*. Amer. Fern Journ. 63(3): 116–127.

— (1974). A revised classification of *Thelypteris* subgenus *Amauropelta*. Amer. Fern Journ. 64(3): 83–95.

— (1975). New species and new combinations of ferns from Chiapas, Mexico. Proc. Calif. Acad. Sci. Ser. 4 40(8): 209–230.

— (1976). New taxa and new combinations of *Thelypteris* from Guatemala. Phytologia 34(3): 231–233.

— (1980). Taxonomy of *Thelypteris* Subgenus *Steiropteris*, including *Glaphyropteris* (*Pteridophyta*). Univ. Calif. Publ. Bot. 76: 1–38.

— (1983). Polypodiaceae – Thelypteroideae. Op. Bot., B (Fl. Ecuador), 18 (14:4): 1–147.

Sota, E.R. de la (1973). Sinopsis de las pteridophytas del noroeste de Argentina II. Darwiniana 18(1–2): 173–263.

Tardieu-Blot, M-L. (1965). A propos de quelques combinations et espèces nouvelles de fougères Africaines ou Malgaches. Adansonia 5(4): 492–502.

Wherry, E.T. (1964). Some new name combinations for Southeastern ferns. Amer. Fern Journ. 54(3): 143–146.